D0848257

Assessment of Mars Science and Mission Priorities

Committee on Planetary and Lunar Exploration
Space Studies Board
Division on Engineering and Physical Sciences

NATIONAL RESEARCH COUNCIL
OF THE NATIONAL ACADEMIES

THE NATIONAL ACADEMIES PRESS
Washington, D.C.
www.nap.edu

THE NATIONAL ACADEMIES PRESS 500 Fifth Street, N.W. WASHINGTON, DC 20001

NOTICE: The project that is the subject of this report was approved by the Governing Board of the National Research Council, whose members are drawn from the councils of the National Academy of Sciences, the National Academy of Engineering, and the Institute of Medicine. The members of the committee responsible for the report were chosen for their special competences and with regard for appropriate balance.

Support for this project was provided by Contracts NASW 96013 and NASW 01001 between the National Academy of Sciences and the National Aeronautics and Space Administration. Any opinions, findings, conclusions, or recommendations expressed in this publication are those of the authors and do not necessarily reflect the views of the sponsor.

Cover: An artist's impression of one of NASA's twin Mars Exploration Rovers on the surface of Mars. Courtesy of NASA and the Jet Propulsion Laboratory.

International Standard Book Number 0-309-08917-4 (Book)
International Standard Book Number 0-309-50833-9 (PDF)

Copies of this report are available free of charge from:

Space Studies Board
National Research Council
The Keck Center of the National Academies
500 Fifth Street, N.W.
Washington, DC 20001

Printed in the United States of America

THE NATIONAL ACADEMIES

Advisers to the Nation on Science, Engineering, and Medicine

The **National Academy of Sciences** is a private, nonprofit, self-perpetuating society of distinguished scholars engaged in scientific and engineering research, dedicated to the furtherance of science and technology and to their use for the general welfare. Upon the authority of the charter granted to it by the Congress in 1863, the Academy has a mandate that requires it to advise the federal government on scientific and technical matters. Dr. Bruce M. Alberts is president of the National Academy of Sciences.

The **National Academy of Engineering** was established in 1964, under the charter of the National Academy of Sciences, as a parallel organization of outstanding engineers. It is autonomous in its administration and in the selection of its members, sharing with the National Academy of Sciences the responsibility for advising the federal government. The National Academy of Engineering also sponsors engineering programs aimed at meeting national needs, encourages education and research, and recognizes the superior achievements of engineers. Dr. Wm. A. Wulf is president of the National Academy of Engineering.

The **Institute of Medicine** was established in 1970 by the National Academy of Sciences to secure the services of eminent members of appropriate professions in the examination of policy matters pertaining to the health of the public. The Institute acts under the responsibility given to the National Academy of Sciences by its congressional charter to be an adviser to the federal government and, upon its own initiative, to identify issues of medical care, research, and education. Dr. Harvey V. Fineberg is president of the Institute of Medicine.

The **National Research Council** was organized by the National Academy of Sciences in 1916 to associate the broad community of science and technology with the Academy's purposes of furthering knowledge and advising the federal government. Functioning in accordance with general policies determined by the Academy, the Council has become the principal operating agency of both the National Academy of Sciences and the National Academy of Engineering in providing services to the government, the public, and the scientific and engineering communities. The Council is administered jointly by both Academies and the Institute of Medicine. Dr. Bruce M. Alberts and Dr. Wm. A. Wulf are chair and vice chair, respectively, of the National Research Council.

www.national-academies.org

OTHER REPORTS OF THE SPACE STUDIES BOARD

Satellite Observations of the Earth's Environment: Accelerating the Transition of Research to Operations (2003)

Assessment of the Usefulness and Availability of NASA's Earth and Space Mission Data (2002)
Factors Affecting the Utilization of the International Space Station for Research in the Biological and Physical Sciences (prepublication) (2002)
Life in the Universe: An Assessment of U.S. and International Programs in Astrobiology (2002)
New Frontiers in the Solar System: An Integrated Exploration Strategy (prepublication) (2002)
Review of NASA's Earth Science Enterprise Applications Program Plan (2002)
"Review of the Redesigned Space Interferometry Mission (SIM)" (2002)
Safe on Mars: Precursor Measurements Necessary to Support Human Operations on the Martian Surface (2002)
The Sun to the Earth—and Beyond: A Decadal Research Strategy in Solar and Space Physics (2002)
Toward New Partnerships in Remote Sensing: Government, the Private Sector, and Earth Science Research (2002)
Using Remote Sensing in State and Local Government: Information for Management and Decision Making (2002)

The Mission of Microgravity and Physical Sciences Research at NASA (2001)
The Quarantine and Certification of Martian Samples (2001)
Readiness Issues Related to Research in the Biological and Physical Sciences on the International Space Station (2001)
"Scientific Assessment of the Descoped Mission Concept for the Next Generation Space Telescope (NGST)" (2001)
Signs of Life: A Report Based on the April 2000 Workshop on Life Detection Techniques (2001)
Transforming Remote Sensing Data into Information and Applications (2001)
U.S. Astronomy and Astrophysics: Managing an Integrated Program (2001)

Assessment of Mission Size Trade-offs for Earth and Space Science Missions (2000)
Ensuring the Climate Record from the NPP and NPOESS Meteorological Satellites (2000)
Future Biotechnology Research on the International Space Station (2000)
Issues in the Integration of Research and Operational Satellite Systems for Climate Research: I. Science and Design (2000)
Issues in the Integration of Research and Operational Satellite Systems for Climate Research: II. Implementation (2000)
Microgravity Research in Support of Technologies for the Human Exploration and Development of Space and Planetary Bodies (2000)
Preventing the Forward Contamination of Europa (2000)
"On Continuing Assessment of Technology Development in NASA's Office of Space Science" (2000)
"On Review of Scientific Aspects of the NASA Triana Mission" (2000)
"On the Space Science Enterprise Draft Strategic Plan" (2000)
Review of NASA's Biomedical Research Program (2000)
Review of NASA's Earth Science Enterprise Research Strategy for 2000-2010 (2000)
The Role of Small Satellites in NASA and NOAA Earth Observation Programs (2000)

Copies of these reports are available free of charge from:
Space Studies Board
The National Academies
500 Fifth Street, NW, Washington, DC 20001
(202) 334-3477
ssb@nas.edu
www.nationalacademies.org/ssb/ssb.html

NOTE: Listed according to year of approval for release.

COMMITTEE ON PLANETARY AND LUNAR EXPLORATION

JOHN A. WOOD, Harvard-Smithsonian Center for Astrophysics, *Chair*
WILLIAM V. BOYNTON, University of Arizona
W. ROGER BUCK, Lamont-Doherty Earth Observatory
JAMES P. FERRIS,[*] Rensselaer Polytechnic Institute
JOHN M. HAYES,[*] Woods Hole Oceanographic Institution
KAREN J. MEECH, University of Hawaii
JOHN F. MUSTARD, Brown University
ANDREW F. NAGY, University of Michigan
KEITH S. NOLL, Space Telescope Science Institute
DAVID A. PAIGE,[*] University of California, Los Angeles
ROBERT T. PAPPALARDO, Brown University
ANNA-LOUISE REYSENBACH, Portland State University
J. WILLIAM SCHOPF, University of California, Los Angeles
ANN L. SPRAGUE, University of Arizona

Staff

DAVID H. SMITH, Study Director
SANDRA J. GRAHAM, Senior Staff Officer
KIRSTEN ARMSTRONG, Research Assistant
BRIAN DEWHURST, Research Assistant
SHARON S. SEAWARD, Senior Program Assistant (through December 2001)
RODNEY N. HOWARD, Senior Program Assistant (from January 2002)

[*]Former member.

Preface

Mars is arguably the most interesting and important target for study in the solar system, because it is the most nearly similar to Earth of all the planets and one of the most likely repositories for extraterrestrial life among them. Its position in the planetary sequence and its relatively benign climate make it more accessible for study than any other planet. Earlier reports and letter reports by the Space Studies Board's Committee on Planetary and Lunar Exploration (COMPLEX)—including *1990 Update to Strategy for Exploration of the Inner Planets* (1990), *An Integrated Strategy for the Planetary Sciences: 1995–2010* (1994), *Review of NASA's Planned Mars Program* (1996), "On NASA's Mars Sample Return Mission Options" (letter report, 1996), and "Assessment of NASA's Mars Exploration Architecture" (letter report, 1998)—have attached a very high priority to Mars exploration, as have the National Aeronautics and Space Administration's own strategic planning documents. The White House National Science and Technology Council's statement of national space policy, dated September 19, 1996, lists as one of the nation's prime space exploration goals "a sustained program to support a robotic presence on the surface of Mars by year 2000 for the purposes of scientific research, exploration and technology development."

The failures in 1999 of two promising Mars missions—Mars Climate Orbiter and Mars Polar Lander—have brought about a critical reexamination and replanning of NASA's program of Mars exploration. It has been several years since COMPLEX last considered its overall scientific priorities for the exploration of Mars, and in this period two very successful missions—Mars Pathfinder and Mars Global Surveyor—have been executed. The discoveries made by these spacecraft, together with related astronomical, theoretical, and laboratory studies, have provided many new data (e.g., detailed maps of the planet's magnetic and gravity fields and topography, history of recent water flow, and composition of surface rocks) that should be included in an assessment of research priorities.

During the current hiatus in activity resulting from NASA's program reassessment, it is appropriate to reexamine the scientific priorities for the exploration of Mars, and it is important to then provide an independent scientific assessment of how well NASA's revised program plans will respond to those scientific priorities. Therefore, in November 2000 the Space Studies Board charged COMPLEX with conducting an assessment of Mars-science and Mars-mission priorities. In particular, the committee was asked to do the following:

- Review the state of knowledge of the planet Mars, with special emphasis on findings of the most recent Mars missions and related research activities;
- Review the most important Mars research opportunities in the immediate future;

- Review scientific priorities for the exploration of Mars identified by COMPLEX (and other scientific advisory groups) and their motivation, and consider the degree to which recent discoveries suggest a reordering of priorities; and
- Assess the congruence between NASA's evolving Mars Exploration Program plan and these recommended priorities, and suggest any adjustments that might be warranted.

Although this project was formally initiated at COMPLEX's January 29–31, 2001, meeting in Tucson, Arizona, the committee heard an extensive series of presentations describing the scientific and technical aspects of Mars exploration during the framing of the charge for this study at COMPLEX's October 2–4, 2000, meeting in Woods Hole, Massachusetts. Work on this project continued at the committee's May 2–4, 2001, meeting in Washington, D.C., and a complete draft of this report was finished in late May 2001. The text was reviewed by the Space Studies Board in June 2001, sent to external reviewers in July 2001, and revised during August and September 2001. Copies of this report were distributed in an unedited, prepublication format in November 2001. This, the final edited text, was prepared in mid-2002 and supersedes all previous versions of this report.

The work of COMPLEX was made easier thanks to the contributions made by Mario H. Acuña (NASA, Goddard Space Flight Center), Raymond E. Arvidson (Washington University), Jay T. Bergstralh (NASA Headquarters), Stephen Bougher (University of Arizona), Bruce A. Campbell (Smithsonian Institution), Stephen M. Clifford (Lunar and Planetary Institute), Peter T. Doran (University of Illinois, Chicago), James Garvin (NASA Headquarters), Martha S. Gilmore (Wesleyan University), James W. Head III (Brown University), G. Scott Hubbard (NASA Headquarters), Bradley L. Jolliff (Washington University), Philippe Masson (University of Paris Sud), and Maria T. Zuber (Massachusetts Institute of Technology).

COMPLEX also wishes to single out for acknowledgment the particularly important contributions made by Victor Baker (University of Arizona) and Mitchell Sogin (Marine Biological Laboratory).

This report has been reviewed by individuals chosen for their diverse perspectives and technical expertise, in accordance with procedures approved by the National Research Council's (NRC's) Report Review Committee. The purpose of this independent review is to provide candid and critical comments that will assist the institution in making the published report as sound as possible and to ensure that the report meets institutional standards for objectivity, evidence, and responsiveness to the study charge. The review comments and draft manuscripts remain confidential to protect the integrity of the deliberative process. COMPLEX thanks the following individuals for their review of this report: James Arnold (University of California, San Diego), Philip R. Christensen (Arizona State University), Alan Delamere (Ball Aerospace), Donald M. Hunten (University of Arizona), Bruce Jakosky (University of Colorado), Norman F. Ness (University of Delaware), and Tobias C. Owen (University of Hawaii). Although the reviewers listed above have provided many constructive comments and suggestions, they were not asked to endorse the conclusions or recommendations, nor did they see the final draft of the report before its release. The review of this report was overseen by Michael Carr (U.S. Geological Survey). Appointed by the National Research Council, he was responsible for making certain that an independent examination of this report was carried out in accordance with institutional procedures and that all review comments were carefully considered. Responsibility for the final content of this report rests solely with the authoring committee and the institution.

Contents

APPENDIXES

Executive Summary

Within the Office of Space Science of the National Aeronautics and Space Administration (NASA) special importance is attached to exploration of the planet Mars, because it is the most like Earth of the planets in the solar system and the place where the first detection of extraterrestrial life seems most likely to be made. The failures in 1999 of two NASA missions—Mars Climate Orbiter and Mars Polar Lander—caused the space agency's program of Mars exploration to be systematically rethought, both technologically and scientifically. A new Mars Exploration Program plan (summarized in Appendix A) was announced in October 2000. The Committee on Planetary and Lunar Exploration (COMPLEX), a standing committee of the Space Studies Board of the National Research Council, was asked to examine the scientific content of this new program. The charge to COMPLEX was as follows:

- Review the state of knowledge of the planet Mars, with special emphasis on findings of the most recent Mars missions and related research activities;
- Review the most important Mars research opportunities in the immediate future;
- Review scientific priorities for the exploration of Mars identified by COMPLEX (and other scientific advisory groups) and their motivation, and consider the degree to which recent discoveries suggest a reordering of priorities; and
- Assess the congruence between NASA's evolving Mars Exploration Program plan and these recommended priorities, and suggest any adjustments that might be warranted.

STUDY APPROACH AND EMPHASIS

COMPLEX comprehensively reviewed Mars science in nine disciplinary areas, working its way from the interior of the planet (Chapter 2) outward to the upper atmosphere (Chapter 10). The committee heard presentations by experts in all these areas and wrote chapters with structures modeled after the charge: Each chapter begins with a review of the present state of knowledge and then discusses near-term opportunities, presents recommended scientific priorities, and offers an assessment of priorities in the Mars Exploration Program.

COMPLEX drew on the publications of 11 earlier committees (including reports from both COMPLEX in years past and NASA) to compile its recommended scientific priorities; these sets of previously published recommendations appear in Appendix B. A document prepared in 2001 by James B. Garvin and Orlando Figueroa of the

NASA Office of Space Science, titled "The Mars Exploration Program: A High-level Description" (reprinted in Appendix A of this report), provided the basis for comparisons with the recommendations, which appear in the section "Assessment of Priorities in the Mars Exploration Program" in each of Chapters 2 through 10. Chapter 12 is a synthesis of the assessments of priorities that appear in Chapters 2 through 10.

COMPLEX judged NASA's responsiveness to the past recommendations of advisory panels to be, on the whole, good, although there is weakness in some areas. COMPLEX acknowledges that budgetary constraints prevent all worthy research goals from being pursued simultaneously, and it endorses the space agency's concentration of effort in areas related to the fundamental question, Did life ever arise on Mars? It is important that efforts to answer this question be broadly based and equally prepared for a negative or positive answer. The implications of either answer to the question will be fully understood only when a broad and deep understanding of Mars has been acquired.

The new Mars Exploration Program recognizes the importance of gaining information about the surface, atmosphere, hydrosphere, and interior of the planet to arrive at an understanding of the Mars dynamic system and a global context for assessment of the biological potential of Mars (see Chapter 12). However, COMPLEX notes that NASA has no plans for missions that address high-priority questions about the interior of Mars. Similarly, there is an absence of NASA missions that specifically address Mars's atmosphere, climate, polar science, ionosphere, and solar wind interactions. Direct measurements of the distribution and behavior of near-surface water are needed. Some but not all of these goals will be addressed by upcoming foreign Mars missions such as Japan's Nozomi, the European Space Agency's Mars Express, the United Kingdom's Beagle 2, and France's NetLander. COMPLEX urges NASA to continue its support for U.S. participation in Mars missions conducted by NASA's international partners. A full picture of the science of Mars is needed to support the quest for martian life.

Water is a topic of particular interest on Mars, not only for its own sake but because it is a prerequisite for life. This is so widely recognized as to have engendered the slogan in advisory and planning circles, "Follow the water." In light of this, there is surprisingly little emphasis on spacecraft experiments designed to detect and study crustal water and ice. COMPLEX believes that this subject needs more attention. In Chapter 12 COMPLEX offers for consideration 10 spacecraft experiments that it considers would advance Mars science, but the committee attempts to be realistic about the factors that militate against them.

COMPLEX attaches special importance to the prospect of sample-return missions to Mars (see Chapter 11). The return of Mars samples has been consistently advocated by advisory panels for more than 20 years as the most effective way to greatly increase our understanding of the history of Mars and its surface environment. No other single strategy can answer so many of the questions about martian chemistry, geology, climatology, and the presence of or potential for life, past or present. Irrespective of the number of orbiters or rovers sent to Mars, we will not come to grips with these fundamental issues until documented samples are available for study in terrestrial laboratories.

The committee believes that enough information is at hand already, much of it from the Mars Global Surveyor mission, to choose the first sites to be sampled intelligently, and that sample return need not wait on additional reconnaissance missions (see Chapters 11 and 12). Even if the first returned samples are not optimal in terms of siting, they will provide a greatly enhanced view of the geologic processes on Mars. Even a "grab sample" of soil from a randomly chosen site on the planet would reveal the character of martian surface material: its chemistry, oxidation state, content of organic materials, mineralogy, and the history of weathering reactions that has affected it. Detailed knowledge of the surface material will permit a more intelligent choice of measurements to be made by future remote-sensing missions. Exercise of selectivity in landing sites will open additional doors: Samples from a formerly fluvial environment, for example, may be found to include rocks of diverse composition and age, including sedimentary rocks that contain a record of aqueous activity on the martian surface, conceivably even fossil evidence of life.

> ***Recommendation.*** Because returned samples will advance Mars science to a new level of understanding, COMPLEX endorses the high priority given to sample return by earlier advisory panels, and it recommends that a sample-return mission be launched at the 2011 launch opportunity.

Since sample-return missions are expensive and will tend to preempt the resources of the Mars Exploration Program, a key question is how many of them are needed. COMPLEX concluded that it cannot be realistically anticipated that the first sample return from Mars will unlock all of the planet's secrets. Orbital observations have shown that Mars's geologic and climatic history is best exposed in widely separated, isolated locations, and a complete picture of martian history is unlikely to be obtained from samples collected at a single location. Instead, the first sample return should be seen as a trailblazer for future sample-return missions, and it should be used to develop the key technologies, procedures, and infrastructure necessary to embark on a future program in which samples are returned from many locations on the planet. COMPLEX estimates that roughly 10 sample-return missions, necessarily executed over a protracted period of time—perhaps three or four decades to as much as a century—may be required to learn the most important things researchers want to know about Mars. A protracted schedule of exploration will allow time for the information gained by each mission to be digested and fed into the planning of the next mission, and it will permit substantial redesign of the spacecraft and sampling system between missions.

> ***Recommendation.*** It should not be anticipated that a few (two to three) Mars sample-return missions will serve the need for samples from that planet. No single site or small number of sites on Mars will answer all of the important questions about the planet, and in any case, the earliest sample-return missions will be in large part technology-development missions. Something like 10 sample-return missions, spread over a substantial period of time, may be required to answer the important questions about Mars.

Considerations of planetary protection dictate that stringent measures be taken during a sample-return mission to prevent biological contamination of either Earth or Mars. One essential measure is the construction of a quarantine facility to receive and contain samples when they arrive on Earth. COMPLEX reiterates the conclusion of its recent report on the design and operation of a Mars Quarantine Facility—that a long lead time is required to prepare such a facility.[a] On the basis of prior experience with facilities of this type, COMPLEX estimated that 7 years will be required to design, construct, and staff the facility. To this period must be added the time needed to clear an environmental impact statement and to carry out several reconnaissance studies that are needed to inform the design and operation of the facility.[b] The aggregate of time required may strain the schedule even of a 2011 launch. The message is plain: Preparations for sample return should not be delayed any longer than they already have been (see Chapter 12).

> ***Recommendation.*** Scientific research and design studies that must precede the design and construction of the Mars Quarantine Facility should begin immediately. Decisions should be made immediately about the siting and management of the facility. Design and construction of the facility should begin at the earliest possible time.

Chapter 7 elaborates on this recommendation, reiterating the conclusions of COMPLEX's recent report on the quarantine of martian samples. In summary, these conclusions are as follows:

- The Mars Quarantine Facility in which Mars samples will be processed, stored, and released for scientific study, and in which a very limited range of studies will be carried out, must be designed, built, and certified.
- Research must be initiated on several outstanding questions that will affect the design of the Mars Quarantine Facility (e.g., combining biological isolation with clean-room conditions; establishing the efficacy and detrimental effects of sterilization techniques).
- The study of life in extreme environments on Earth, which can aid in the design of life-detection tests, should be supported, as is already being done. In general, research areas that improve the sensitivity of life-detection techniques must be supported, and a life-detection protocol to be implemented and tested in the Mars Quarantine Facility must be developed.
- Techniques must be developed for the collection, packaging, and return of samples.

---

[a]Space Studies Board, National Research Council, *The Quarantine and Certification of Martian Samples*, National Academy Press, Washington, D.C., 2002.

[b]For more details, see Space Studies Board, National Research Council, *The Quarantine and Certification of Martian Samples*, National Academy Press, Washington, D.C., 2002.

- The research programs of Mars orbiter and lander missions must be designed to support the collection of those samples with the greatest potential for life detection (this process is underway).

ADDITIONAL CONCERNS

COMPLEX also considered several other topics important to Mars exploration (see Chapter 12), including those addressed below.

Power Supplies for Landers and Rovers

An extremely important consideration in establishing the capabilities of landed packages, static or roving, on Mars is the power supply on which they rely—the options being solar panels and radioisotope power systems. The limitations placed on landed spacecraft by solar power supplies are reviewed. From a science viewpoint, the advantages of nuclear power—namely, long-lived missions, night-time operation, and access to any point on Mars—are clear. COMPLEX urges the use of radioisotope power systems, if at all feasible, on advanced Mars lander missions.

The Mars Scout Program

There is concern in the scientific community that the Mars Scout Program, the youngest and smallest element of the Mars Exploration Program, may also be the most vulnerable. The fear is that the Scout Program may not achieve its potential because it will be sacrificed in times of budget stringency. COMPLEX considers that this would be unfortunate.

Recommendation. So that the Mars Scout missions can fulfill their laudable goals of filling in gaps in the Mars Exploration Program and allowing a rapid response to scientific discoveries, COMPLEX recommends that care be taken to maintain the Mars Scout Program as a viable line of missions when budget problems arise.

Data Analysis, Ground-Based Observations, and Laboratory Analysis

COMPLEX also noted that the Mars Exploration Program, with its missions at 2-year intervals, presents a new problem in fully exploiting the amount and variety of data that will be collected. The volume and quality of data returned by Mars Global Surveyor alone have been extraordinary, and the analysis of these data is only beginning. With the rapid pace of Mars missions planned for the next decade, the flood of data can be expected to increase. This problem should be recognized, and NASA's data analysis and science programs should be structured to accommodate and support the broad range of Mars science that is to come.

Recommendation. A plan should be developed at the program level, not at the level of each mission, for archiving and making accessible the data to be gathered by the Mars Exploration Program. It is essential that support be provided for the study and exploitation of this body of data.

The Mars Exploration Program consists of a queue of flight missions, so the present assessment of the program also discusses flight missions and rarely touches on Earth-based research. However, the latter is an essential component of the total program of Mars research, and in Chapter 1 of this report, COMPLEX acknowledges several areas of Earth-based Mars research, urging continued support of these and other areas of Earth-based research, because they are essential to a balanced program of Mars research.

Recommendation. COMPLEX endorses continued support for nonflight activities such as ground-based observing and laboratory analysis.

1

Introduction

Mars occupies a special place in the U.S. program of space exploration, and also in the minds of the public, because it is the most Earth-like planet in the solar system and the place where the first detection of extraterrestrial life seems most likely to be made. Another important reason for studying Mars is to be able to compare it with the other terrestrial planets and to learn how the differences among these planets relate to differences in their formation and evolution, stemming from factors such as their distances from the Sun, their initial sizes, and their proximity to Jupiter.

The Mars Surveyor Program, begun in 1996 after a 20-year hiatus in (successful) U.S. Mars missions, was to be an ambitious exploration of the Red Planet, inspired by the success of the modestly supported Pathfinder Lander mission in that year, and also by reports that the martian meteorite designated ALH84001 contains possible evidence of extraterrestrial life. However, with the failures of the Mars Climate Orbiter and Mars Polar Lander missions in 1999, the Surveyor program came to be seen as unworkable, and in the time since those failures the strategy for Mars exploration has been systematically rethought and its more ambitious goals have been scaled back. A new Mars Exploration Program (MEP)—no longer the Mars Surveyor Program—has been developed, building on earlier missions and on the success of an ongoing orbital mission, Mars Global Surveyor (MGS), which has already returned a wealth of data that have revolutionized our understanding of the planet. Missions to Mars will be launched at every launch opportunity, i.e., on approximately 26-month centers.

The new Mars Exploration Program was announced by NASA's Office of Space Science on October 26, 2000. (The MEP is described in Appendix A of this report.) The task of the present study was to assess the program in the light of recommendations made earlier to NASA by the Committee on Planetary and Lunar Exploration (COMPLEX) and other advisory panels (these recommendations are summarized in Appendix B of this report) and to consider the degree to which recent discoveries suggest a reordering of priorities.

The task of reviewing Mars science is daunting. An exhaustive summary would be far beyond the scope of National Research Council (NRC) reports, and COMPLEX is quick to admit that this report is far from exhaustive. The report first reviews nine topics that comprehensively describe contemporary Mars science (Chapters 2 through 10), working from the interior of the planet outward. Each chapter summarizes recommendations that have been made relative to that topic. Section numbers in square brackets (e.g., [1.10]) reference specific recommendations that appear in Appendix B. Rather than discussing a topic of Mars science, Chapter 11 addresses a technique, and a particularly important one: the collection and return of samples from Mars. Chapter 12 then presents a summary discussion of priorities in the Mars Exploration Program in the light of earlier recommendations and current realities, and Chapter 13 concludes the report. A key to the labyrinth of acronyms used in space exploration is included as Appendix C.

This study does not treat the satellites of Mars—Phobos and Deimos—as these have not been a quest of the NASA flight program, and their science is rather far removed from that of the terrestrial planets, being more closely aligned with that of the asteroids.

The Mars Exploration Program consists of flight missions, and the task of the present study is to discuss the science and priorities connected with flight missions. However, the exploration of Mars involves many modes of data acquisition and scientific inquiry, and it is important to keep in mind the essential elements of Mars science that stem from Earth-based research. These include not only the analysis of data from flight missions, without which the data themselves would be useless, but also purely ground-based research: telescopic studies, theoretical modeling and analysis, and a variety of studies in terrestrial laboratories. Because this report is directed toward an assessment of the Mars Exploration Program, it rarely addresses the important science carried out in Earth-based studies, but these must be considered an important part of any integrated scientific exploration. Some examples follow:

- *Telescopic studies.* Continuing telescopic observation of Mars (Figure 1.1) has played a key role in demonstrating that the surface of Mars changes on a relatively short time scale (examples of such changes include seasonal cycles, dust storms, and evolution of the polar caps). Telescopic and spacecraft data are highly synergistic, and each type plays a role in supporting the other. NASA's Infrared Telescope Facility, a 3-m infrared-optimized telescope on Mauna Kea, is an important facility for planetary science because it can be dedicated to mission support.[1] Near-infrared spectra can be used to distinguish between water and CO_2 ice clouds,[2] and the 4.6-μm CO band can be used to monitor the dust content of the atmosphere. Atmospheric water vapor has been observed remotely[3,4,5] and in situ by the Viking[6,7] and Pathfinder[8] missions. Support for future robotic and possible manned missions to Mars will require a long climatological baseline. The long baseline, partially obtained with ground-based and Hubble Space Telescope data, will also contribute to an understanding of the water cycles between the atmosphere, regolith, and polar caps, as well as providing spatially resolved data on volatile cycles of H_2O, CO_2, CO, and O_3.
- *Theoretical models.* Models are an essential component of any scientific endeavor. Examples of theoretical planetary studies are those that treat the geodynamics of Mars, its interior structure, atmospheric loss and fractionation,[9] and global climate and general circulation models. Climate models, which are currently adapted from terrestrial general circulation models, are becoming increasingly important, yet they will require much additional observational data, particularly of surface-atmosphere energy and gas fluxes, for model validation and verification.[10] The geodynamical investigations often study controls on the obliquity of Mars, and the variability of that parameter, which is so crucial to the planet's climate history and the prospects of life on it.[11] Theoretical studies of the interior attempt to model Mars's core, the composition and viscous behavior of the mantle (the latter controls the tectonics of martian crust), and the magnetic record in the planet.[12]
- *Martian meteorites.* Evidence is very strong that the SNC[a] category of meteorites is cratering debris from Mars. Studies of this small group of meteorites in terrestrial laboratories have provided invaluable, if fragmentary, information about the geochemistry and chronology of the planet (see Chapter 3).[13,14] Five of the SNC meteorites being studied were collected from the antarctic ice sheet by teams of searchers supported by NASA, the National Science Foundation, and the Smithsonian Institution (others are finds from deserts, or observed falls). The antarctic meteorite program was instituted in 1976; under this program, teams of experts search areas known to contain a concentration of meteorites for 6 weeks every austral summer. The research and analysis that led to the conclusion that SNC meteorites come from Mars is an excellent example of how support in the basic research of planetary materials has contributed significantly to our understanding of the planet.

[a]The acronym comes from the three classes of meteorites in this category—shergottites, nakhlites, and chassignites.

FIGURE 1.1 Hubble Space Telescope's Wide Field and Planetary Camera 2 took these images of Mars between April 27 and May 6, 1999, when Mars was 87 million kilometers from Earth. Together, the four images show the entire martian surface, upon which features as small as 19 kilometers across are visible. NASA image courtesy of Steven Lee (University of Colorado), James Bell (Cornell University), and Michael Wolff (Space Science Institute).

- *Astrobiological research.* Studies of Earth's deep-sea hydrothermal environments, hot springs, the deep subsurface, alkaline or acidic environments, and sea ice have revealed amazing microbial diversity in the form of uncultured organisms from environmental extremes. Some of these habitats are analogous to past and present martian environments where life may have arisen or might continue to exist. The search for living organisms is no longer constrained by a requirement for photosynthesis. Microbial species capable of subsurface growth in the presence of high concentrations of metals, high and low pH, and in either extremely cold or hot conditions are known. Despite discoveries of environmental extremes compatible with life, we have only limited knowledge of microbial diversity, the conditions under which such species live, and how interactions between microbial forms modulate planetary change. Ground-based astrobiology supplies a clear rationale and direction for the selection of landing sites on Mars, allows for the proper design and interpretation of in situ experiments, and provides the basis for life detection and planetary protection. It is imperative to develop sensitive life-detection protocols that will not be confused by terrestrial contamination; we must establish effective means for sterilizing returned samples without compromising their value for nonbiological studies; and through expanded knowledge about potential diversity of the microbial world, we must explore how ancient microbial life might have affected planetary processes on Mars. Through these investigations, we will be positioned to optimize information from Mars in situ and sample-return missions.
- *Other laboratory studies.* Inputs into theoretical studies, modeling, instrument design, and spacecraft missions are in part derived from terrestrial laboratory studies. In these, basic measurements are made of chemical

reaction rates, absorption cross sections, scattering cross sections, and other parameters that are important to studies of the martian surface and atmosphere and understanding of processes in them.[15]

COMPLEX stresses that continued support of these and other areas of Earth-based research is essential to a balanced program of Mars research (see Chapter 12 in this report).

REFERENCES

1. M.S. Hanner, K.J. Meech, E. Barker, M.J.S. Belton, R. Binzel, and J. Spencer, *The Future Role of the IRTF—Report to NASA from the NASA IRTF/Keck Management Operations Working Group*, 1998.
2. See, for example, D.R. Klassen, J.F. Bell III, R.R. Howell, P.E. Johnson, W. Golisch, C.D. Kaminski, and D. Griep, "Infrared Spectral Imaging of Martian Clouds and Ices," *Icarus* 138: 36–48, 1999.
3. E.S. Barker, "Martian Atmospheric Water Vapor Observations: 1972–74 Apparition," *Icarus* 28: 247–268, 1976.
4. B. Rizk, R.M. Haberle, D.M. Hunten, and J.B. Pollack, "Meridional Transport and Water-Reservoirs in Southern Mars During 1988–1989," *Icarus* 118: 39–50, 1995.
5. A.L. Sprague, D.M. Hunten, R.E. Hill, L.R. Doose, and B. Rizk, "Martian Atmospheric Water Abundances: 1996–1999," *Bulletin of the American Astronomical Society* 32: 1093, 2000.
6. C.B. Farmer, D.W. Davies, A.L. Holland, D.D. La Porte, and P.E. Doms, "Mars: Water Vapor Observations from the Viking Orbiters," *Journal of Geophysical Research* 82: 4225–4248, 1977.
7. B.M. Jakosky, and C.B. Farmer, "The Seasonal and Global Behavior of Water Vapor in the Mars Atmosphere: Complete Global Results of the Viking Atmospheric Water Detector Experiment," *Journal of Geophysical Research* 87: 2999–3019, 1982.
8. D.V. Titov, W.J. Markiewicz, N. Thomas, H.U. Keller, R.M. Sablotny, M.G. Tomasko, M.T. Lemmon, and P.H. Smith, "Measurements of the Atmospheric Water Vapor on Mars by the Imager for Mars Pathfinder," *Journal of Geophysical Research* 104: 9019–9026, 1999.
9. See, for example, D.M. Hunten, R.O. Pepin, and T.C. Owen, "Elemental Fractionation Patterns in Planetary Atmospheres," in *Meteorites and the Early Solar System*, J. Kerridge and M.S. Matthews (eds.), University of Arizona Press, Tucson, 1988, pp. 565–591.
10. See, for example, R.H. Haberle, "Early Mars Climate Models," *Journal of Geophysical Research* 103: 28467–28480, 1998; and S.W. Bougher, S. Engel, R.G. Roble, and B. Foster, "Comparative Planet Thermospheres: 3. Solar Cycle Variation of Global Structure and Winds at Solstices," *Journal of Geophysical Research* 105: 17669–17692, 2000.
11. See, for example, G. Spada, and L. Alfonsi, "Obliquity Variations Due to Climate Friction on Mars: Darwin Versus Layered Models," *Journal of Geophysical Research* 103: 28599–28606, 1998; and B.G. Bills, "Obliquity-Oblateness Feedback on Mars," *Journal of Geophysical Research* 104: 30773–30798, 1999.
12. See, for example, C.L. Johnson, S.C. Solomon, J.W. Head, R.J. Phillips, D.E. Smith, and M.T. Zuber, "Lithospheric Loading by the Northern Polar Cap on Mars," *Icarus* 144: 313–328, 2000; K.F. Sprenke and L.L. Baker, "Magnetization, Paleomagnetic Poles, and Polar Wander on Mars," *Icarus* 147: 26–34, 2000; and P. Defraigne, V. Dehant, and T. Van Hoolst, "Steady-State Convection in Mars' Mantle," *Planetary and Space Science* 49: 501–509, 2001.
13. R.C. Wiens, R.H. Becker, and R.O. Pepin, "The Case for Martian Origin of the Shergottites: II. Trapped and Indigenous Gas Components in EETA 79001 Glass," *Earth and Planetary Science Letters* 77: 149–158, 1986.
14. H.Y. McSween, "What We Have Learned About Mars from SNC Meteorites," *Meteoritics* 29: 757–779, 1994.
15. See, for example, D. Kella, P.J. Johnson, H.B. Pedersen, L. Vejby-Christensen, and L.H. Andersen, "The Source of Green Light Emission Determined from a Heavy-Ion Storage Ring Experiment," *Science* 276: 1530–1533, 1997; and E.S. Hwang, R.A. Bergman, R.A. Copeland, and T.G. Slanger, "Temperature Dependence of the Collisional Removal of O_2 ($b^1\Sigma_g^+$, $v = 1$ and 2) at 110–260 K, and Atmospheric Applications," *Journal of Chemical Physics* 110: 18–24, 1999.

2

Interior and Crustal Structure and Activity

PRESENT STATE OF KNOWLEDGE

Major advances in our understanding of the interior and crustal structure of Mars have come recently in four important areas:

1. The bulk composition of Mars is better constrained as a result of the vastly improved estimate of the moment of inertia, made possible by Pathfinder measurements.[1]
2. Mars had a magnetic field in the past, but there is no present global field, as shown by high-amplitude magnetic anomalies detected in the southern highlands of Mars by the Mars Global Surveyor (MGS).[2]
3. An impact origin for the hemispheric dichotomy between the northern and southern hemispheres of Mars appears to be ruled out by topography and gravity data from MGS.[3]
4. Links between a warm, wet climatic period and the formation of the gigantic Tharsis Plateau are suggested by the MGS topography and gravity data.[4]

Before targets for future inquiry are described, these four subjects are discussed briefly.

Bulk Composition

Constraining Mars's bulk composition is an excellent way to improve our insight into processes of planet formation in the solar system.[5] The composition of a planet is constrained by measurements of its mean density (well determined for the inner planets), the moment-of-inertia factor,[a] the seismic velocity structure, and from the composition of partial melts (now rocks) extruded from the interior. The moment of inertia depends on the distribution of density within a planet, and only a limited range of rock compositions have a given density.

Before the Mars Pathfinder mission in 1996, the moment of inertia was known with certainty only for Earth and the Moon. The large uncertainty previously associated with the moment of inertia of Mars, due to an uncertainty in the spin pole precession rate, left room for a range of feasible bulk composition models. Folkner and

[a]The moment-of-inertia factor (K) of a body, a dimensionless number, equals the body's moment of inertia divided by its mass and the square of its radius. K is descriptive of the distribution of mass within the body. $K = 0.4$ for a body with uniform density; the value gets smaller as mass is increasingly concentrated near its center. $K = 0.3315$ for Earth.

colleagues calculated a value of 0.3662 ± 0.0017 for the moment-of-inertia factor, based on an improved estimate of the martian spin pole precession rate determined from Doppler range measurements to the Mars Pathfinder and Viking landers.[6]

Any compositional model of Mars must account for its low mean density and high moment-of-inertia factor compared with those of Earth. Uncompressed density is a more relevant indicator of planetary composition, but this is necessarily uncertain because of a lack of knowledge of temperatures in the planet and material properties at high pressure. The ranges of estimated uncompressed densities are 3.8 to 3.9 g/cm^3 for Mars and 4.4 to 4.5 g/cm^3 for Earth.[7] The moment-of-inertia factor measured for Earth is 0.3315.[8] The difference in moment-of-inertia factors indicates that the concentration of mass toward the center of Mars is less than that of Earth. Earth's core size is well determined from seismic studies; thus, the measured moment-of-inertia factor for Earth directly constrains the ratio of the core density to the mantle density. However, on Mars the core size is still a free parameter. As shown in Figure 2.1, the moment-of-inertia factor and bulk density constraints allow a wide range of possible core sizes, with an attendant range of core and mantle compositions.

Two factors are thought to be responsible for the lower mean density of Mars compared with that of Earth.[9] Mars has an Fe/Si ratio smaller than that of Earth, and it is somewhat enriched in oxygen. The additional O is chiefly in the mantle, associated with Fe^{2+} that would otherwise occur as Fe^0 in the core. A smaller or lighter core alone would not account completely for the higher moment-of-inertia factor of Mars: It is also necessary that the average density of Mars's mantle be greater than that of Earth's mantle. No unique density can be specified by the geophysical constraints, however, since this value depends on assumptions of crustal thickness and core size. The

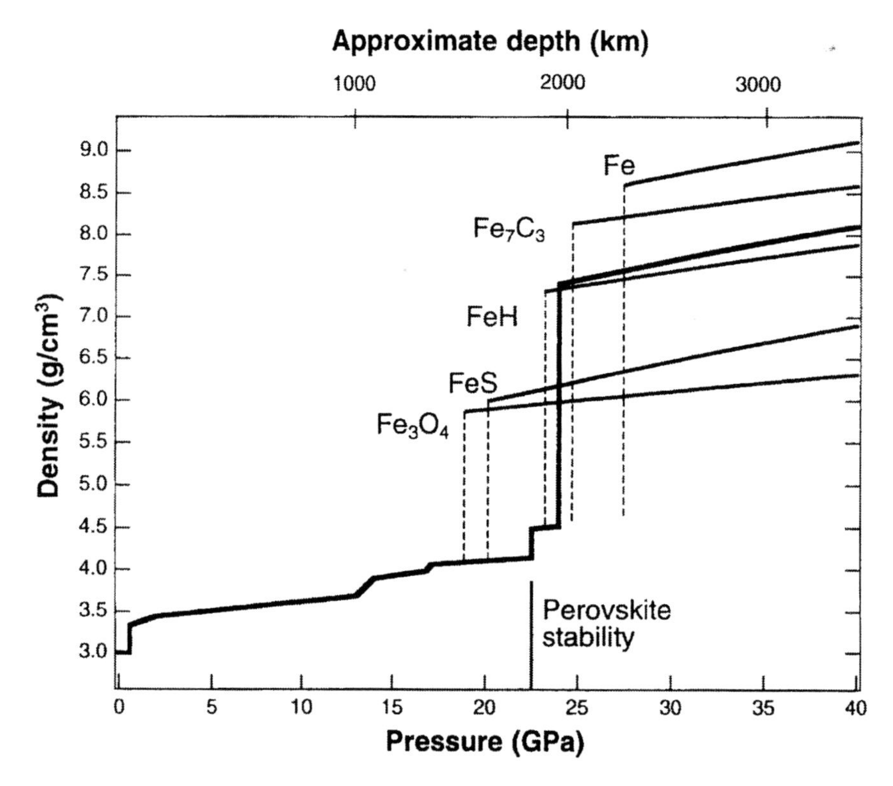

FIGURE 2.1 Density profiles for a range of model core compositions (solid lines). For each core composition, the thickness of the low-density crust is adjusted to give the correct mean density and moment of inertia for Mars. Dashed lines indicate the depth, or pressure, of the core/mantle boundary for model core compositions. The crust-mantle-and-core profile shown (heavy line) assumes a 50-km, 3.0-g/cm^3 crust. SOURCE: Reprinted with permission from C.M. Bertka and Y. Fei, "Implications of Mars Pathfinder Data for the Accretion History of the Terrestrial Planets," *Science* 281: 1838–1840, 1998. Copyright 1998 by the American Association for the Advancement of Science.

upper and lower bounds for mantle density are about 3.45 and 3.55 g/cm^3.[10,11,12] The uncompressed density of material likely to be in Earth's upper mantle is about 3.34 g/cm^3.[13]

Given that Mars must contain less Fe and more O than do the planets closer to the Sun, there are several ways to accomplish this within reasonable cosmochemical frameworks.[14,15] The Equilibrium Condensation Model assumes that the temperatures and pressures in the zone of accretion of Mars were low enough to promote condensation of a different suite of minerals there than in the zones of the other terrestrial planets.[16] Low temperature would allow more complete reaction of S with Fe to form FeS, affecting core density; and more Fe would oxidize and form ferrous silicates, resulting in an increased mantle density. Ringwood and Clark have assumed that Mars, Earth, and Venus have the same abundances of the common metals Fe, Mg, Si, and Al, but differing amounts of O.[17]

Zones of Strong Crustal Magnetization and No Global Field

During the aerobraking phase of the MGS mission the orbiter flew below the martian ionosphere and the onboard magnetometer detected regions with very-high-amplitude magnetic anomalies.[18] The short (~100-km) wavelength of these anomalies implies that the source of the field is confined to the top several tens of kilometers of the crust. The estimated crustal magnetizations are in the range of 10 to 30 A/m, an order of magnitude stronger than magnetizations typically encountered for Earth rocks.[19] Over much of the planet, the pattern of the anomalies is blocky or mottled. The field seen in parts of the highlands is consistent with a crustal magnetization model made up of multiple quasi-parallel linear features with dimensions ~200 km in width, extending as far as 2000 km in length. Some workers interpret these highland areas as remnants of early "oceanic crust," reworked in places by subsequent major impacts and thermal events, but preserving elsewhere the magnetic imprint acquired when the crust formed by "sea floor spreading."[20] Others consider that dike intrusions produced the magnetization patterns.[21]

Measurements made early in the MGS mission established unambiguously that Mars does not currently possess a significant global magnetic field, the estimated upper limit for a Mars dipole moment being $\sim 2 \times 10^{18}$ A m^2.[22] The absence of crustal magnetism near large impact basins such as Hellas and Argyre implies cessation of internal dynamo action during the early Noachian epoch (~4 billion years ago).

Crustal Thickness and the Hemispheric Age Dichotomy

The northern hemisphere of Mars is far smoother and less cratered than is the southern hemisphere, and the northern hemisphere is about 5 km lower in elevation than the south.[23] Ideas for the origin of the dichotomy range from impact thinning of the crust[24,25] to formation of thin basaltic crust in some form of plate tectonics early in Mars's history.[26,27] The smoothness of the lowland surface is attributed to either volcanic or sedimentary infilling. The great number of erosional channels linking the highlands to the lowlands favors sedimentary infill.[28]

The topography and gravity data acquired with the Mars Global Surveyor spacecraft have allowed estimation of the range of variability of crustal thickness over the planet.[29] Crustal thickness variations are fairly smooth across the dichotomy boundary (see Figure 2.2); thus, an impact origin for the lowlands is not favored. The crust thickness results are not inconsistent with the plate tectonic hypothesis but certainly do not confirm that idea.

The gravity and topography analysis also allows estimates of a lower limit for the average crustal thickness, for the case in which the crust is assumed to approach zero thickness at its thinnest point. This yields an estimate of 50 km as the minimum average crustal thickness over the planet.[30] Studies of the geochemistry of martian meteorites suggest that the amount of crustal material differentiated from the mantle would produce a crust thicker than 100 km.[31] Geophysical upper bounds on the crustal thickness are less well defined, since they depend on arguments about the temperature-dependent flow strength of the crust and the temperature of the lower crust. Basically, the thicker the crust, the hotter and therefore weaker is the lower crust. Assuming a temperature gradient and crustal thickness, the temperature can be estimated, and from it the strength of the lower crust. If the lower crust is too hot, it would be expected to flow and in so doing to obliterate differences in topography. The fact that the southern hemisphere is about 5 km higher than the northern hemisphere indicates that the crust is too cold to

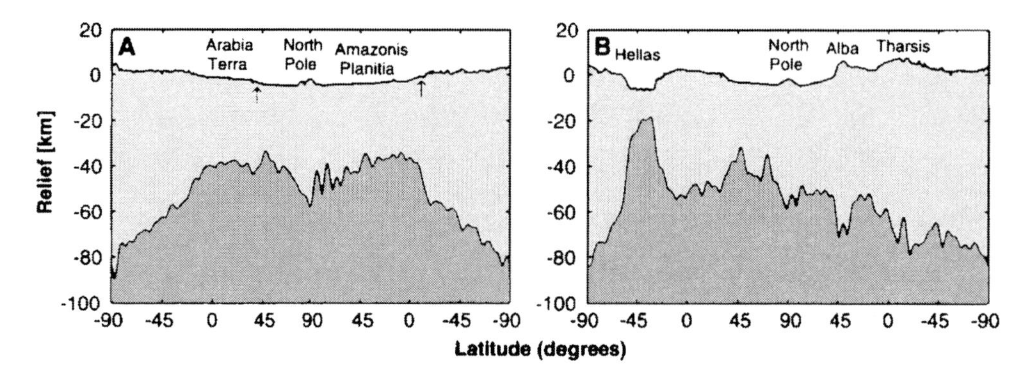

FIGURE 2.2 Circum-Mars profiles of crustal thickness along longitude lines of (A) 0° to 180° E and (B) 70° to 250° E. Light gray represents crust, and dark gray represents mantle. In the figures, the south pole is at both ends of the plot, the north pole is at the center, and the lower-longitude profiles (0° E and 70° E) are on the left sides of the plots. Apparent crustal thickening beneath the north and south polar regions is an artifact of the assumption that layered terrains and ice caps are composed of material with the same density as the crust rather than of less dense ice plus dust. The arrows in (A) show the location of the hemispheric dichotomy boundary. The vertical exaggeration is 30:1. SOURCE: Reprinted with permission from M.T. Zuber, S.C. Solomon, R.J. Phillips, D.E. Smith, G.L. Tyler, O. Aharonson, G. Balmino, W.B. Banerdt, J.W. Head, C.L. Johnson, F.G. Lemoine, P.J. McGovern, G.A. Neumann, D.D. Rowlands, and S.J. Zhong, "Internal Structure and Early Thermal Evolution of Mars from Mars Global Surveyor Topography and Gravity," *Science* 287: 1788–1793, 2000. Copyright 2000 by the American Association for the Advancement of Science.

flow now and was too cold to flow when the hemisphere difference was created. If the interior had been extremely hot at the time of formation of the large impact basins, such as Hellas, then the topographic relief of those features would not have been preserved.[32]

Climate, Lithosphere, Volcanoes, and Support of the Tharsis Plateau

The high-elevation Tharsis region is the main area where tectonic faults and volcanoes are exposed. The surface of the region is young because of volcanic resurfacing, although there are no absolute dates on the age of those volcanoes. Evidence of very recent Tharsis volcanism is indirect and comes from basaltic meteorites from Mars, the shergottites, which have igneous crystallization ages of ~180 million years.[33] Small offset faults appear to be related to lithospheric bending in response to the loading of huge volcanoes. Normal faults also parallel the huge Valles Marineris canyon system, lending weight to the idea that this is a rift.

Two different mechanisms have been proposed to explain the long-wavelength gravity anomalies associated with the Tharsis Rise. First, the gravity anomalies are explained as a result of the load of Tharsis volcanic rocks on an ~100-km-thick elastic lithosphere with insignificant contribution from the deep interior.[34,35] Second, the gravity and topography of the Tharsis Rise are interpreted as dynamic effects of mantle convection.[36]

A key insight from the MGS topographic data is that the plateau predates the formation of apparently fluvial channels. This suggests that the outpouring of lava to make the plateau may have released enough carbon dioxide to form an insulating atmosphere and sufficient water to form the channels and even an ocean. Tharsis loading amounts to ~3×10^8 km^3 of igneous material, which is equivalent to a 2-km-thick global layer. Assuming that the Tharsis magmas have a volatile content similar to that of Hawaiian basalts, the total release of gases from Tharsis magmas could produce the integrated equivalent of a 1.5-bar CO_2 atmosphere and a 120-m-thick global layer of water.[37] These quantities of volatiles may have been sufficient to warm the atmosphere to the point at which liquid water is stable at the surface, which could explain the evidence for water-cut channels around Mars.[38] This is a particularly clear example of how integrating basic geophysical and geological measurements has led to inferences about the past climate of the planet.

NEAR-TERM OPPORTUNITIES

Two near-term launch opportunities would address high-priority scientific objectives relating to the structure of Mars. The Mars Reconnaissance Orbiter, intended for a 2005 launch, is to include gravity mapping at a higher resolution than has been done by MGS. This will allow improved estimates of the near-surface density and crustal thickness variations. Combined analysis of higher-resolution gravity and MOLA (Mars Orbiter Laser Altimeter on MGS) topography will lead to better estimates of the lithospheric thickness under volcanoes, rifts, and craters. If the crustal thickness in a few places can be determined seismically, then the gravity data can be used to quantify the crustal thickness relative to those places.

A passive seismic experiment that might determine crustal thickness in one place is in the advanced planning stage: the NetLander project, planned and supported by several national space agencies in Europe—including those of France, Germany, Finland, and Italy—with some U.S. participation, is envisioned to include four broadband seismic stations. Though this planned experiment involves a minimal number of stations, it may accomplish several important goals, including constraining the size and state of the core, and determining the level of tectonic activity and the rate of meteroid impacts. This is an excellent start to the exploration of Mars's interior, and COMPLEX encourages NASA to support this and more comprehensive future seismic experiments.

RECOMMENDED SCIENTIFIC PRIORITIES

Virtually everything about Mars, including its surface composition, topographic relief, and even its atmosphere, has its origin in processes of the planet's interior. Thus, many kinds of studies would shed light on the composition and evolution of the interior. For example, multispectral analysis from orbit will help constrain the composition of surface rocks and so give a better idea of the kind of interior from which the rocks came. Clearly, analysis of returned rock samples will improve our knowledge of the source region of those rocks in the same way that samples from the Moon did. Age-dating of rocks on the surface could indicate when the major tectonic features of the planet formed and so constrain its thermal evolution. Since each of these types of measurement is central to objectives discussed in later chapters of this report, they are not addressed further in this section—which instead focuses on geophysical measurements that are crucial to major advances in our knowledge of the interior of Mars. Among the types of measurements to address scientific objectives listed below, the passive seismic experiment has by far the highest priority.

Deep Internal Structure

The most direct way to constrain the deep structure of Mars is by use of seismic waves generated by activity within the planet and detected by an array of seismometers distributed across the planet's surface (Appendix B: [1.11]).

There are several clear and compelling reasons for deploying an array of broadband seismometers on Mars. First, seismic data can determine the size of the core (Appendix B: [11.1.7, 3.1]). When the size of the core is known, there can be much tighter constraints on the bulk composition of the planet. Information on the seismic velocities in the mantle will also constrain its composition. This will provide the best measure of the composition of a terrestrial planet outside the Earth-Moon system, and so constrain cosmochemical models for the origin of planetary systems (Appendix B: [1.2, 4.7, 10.1]).

Second, seismology will tell us whether the interior of Mars is active today. If evidence of activity were found, it would generate great interest in further studying a planet that is not dead (Appendix B: [1.11]). As noted above, there is evidence from some Mars meteorites that Mars may have been volcanically active in the geologically recent past.

Third, the question of whether the core is all solid, all liquid, or part solid and part liquid (as is the Earth's core) has a direct and profound bearing on our understanding of planetary dynamos and the present-day lack of a Mars global magnetic field (e.g., Appendix B: [11.3.5]).[39] Several groups have concluded that a passive seismometer experiment could answer these questions, in that there would likely be enough natural sources to generate the needed seismograms (e.g., Appendix B: [1.11]).[40] (The one Viking seismometer that deployed did not detect Mars quakes, but it was not well coupled to the ground and it had a sensitivity as much as four orders of magnitude less than it is possible to use now.[41,42]) The issue of fluid in the core can also be constrained by tracking the positions of two or more landers for rotational dynamics (Appendix B: [1.12, 4.1]).

Shallow Seismic Structure

One key question is, How thick is the crust of Mars? (Appendix B: [1.12, 3.1, 11.6.3]). With the thickness and velocity profile through the crust determined at several locations, models could be constrained for the origin of the hemispheric dichotomy, thus improving our understanding of the geologic history of Mars. Receiver function analysis of passive seismic stations could constrain the crustal thickness and shallow mantle structure.

Active seismic experiments could elucidate the smaller-scale problems of the structure of the crust and shallow mantle. Local- or regional-scale active experiments could also help determine the thickness and stratigraphy of sedimentary and volcanic units, as well as differentiating between those units. Such experiments require a far higher level of effort to carry out than do passive experiments.

Heat Flow

One of the most basic pieces of knowledge needed is whether the interior of Mars is currently hot enough to flow, melt, and cause tectonic activity (Appendix B: [11.3.5, 11.7.1]). On Earth, hydrothermal activity and associated life forms are concentrated in areas of tectonic and magmatic activity. Is there such activity on Mars? One way to begin to answer this is to carry out heat flow measurements on the surface in a wide range of environments (a form of the geophysical mapping recommended in Appendix B: [1.18]). These measurements will require drilling into the surface to emplace heat flow probes, as was done for the Apollo measurements on the Moon.

Gravity Field

Gravity field measurements on a finer scale than accomplished by MGS can help constrain the subsurface structure of sedimentary basins and variations in crustal thickness (Appendix B: [1.11, 1.18, 11.7.1]). For structures on a scale less than the order of 100 km, such measurements probably cannot be made by an orbiter.

Magnetic Field

There is great interest in the origin of the very strong remnant magnetic anomalies found by MGS and discussed above. To learn more about these features, researchers need to map magnetic anomalies on a finer scale than has been done so far. There are several ways to improve the spatial resolution of the magnetic field maps, including orbiter missions with aerobraking, near-surface measurements (either by rovers or aircraft), and sampling and study of the magnetic properties of rocks. Inclusion of a magnetometer in future orbiter missions using aerobraking would bring the instrument close to parts of the martian surface that were not mapped at periapses (closest approach) during the MGS mission (Appendix B: [3.1, 11.7.1]).

ASSESSMENT OF PRIORITIES IN THE MARS EXPLORATION PROGRAM

The top-priority objectives concerning the interior of Mars are *not* high priorities in the first 10 years of NASA's new Mars Exploration Program. The European NetLander mission, which is being planned in conjunction with NASA, does focus on the highest-priority goals, and COMPLEX applauds this effort. There are opportunities for gravity, heat flow, and magnetic measurements on several missions that are expected to fly in the next decade, and COMPLEX believes that they should be pursued. It is critical to understand the structure and evolution of the planet as a whole in order to understand discoveries relating to distributions of water and conditions that might have allowed life to develop.

Great improvement in our knowledge of the martian interior could come from a passive seismic experiment. There are good reasons, based on terrestrial and lunar experience, to believe that the rate of detectable seismicity on Mars would provide adequate sources for imaging the interior. The impacts of meteoroids would also provide useful seismic signals. Determining the rate, mechanism, and location of seismicity and the rate of meteoroid impacts are high-priority objectives in themselves. Crustal structure in a single area could potentially be constrained with a local network of three stations. Seismic attenuation properties should depend on whether the shallow crust is wet (as on Earth) or dry (as on the Moon). More stations would be needed to locate Mars quakes accurately and to determine the velocity and attenuation structure of the mantle and the size and state of the core. For example, accurate constraints on the size and state of the core would require stations at a wide range of distances from seismic sources. Thus, a phased approach is suggested, with a deployment of four stations to determine the locations of seismic sources and constrain crustal properties, while later experiments would add more stations to focus on the deep interior.

The seismic experiment part of the NetLander mission would take the first step in a phased seismic study of the martian interior. It must, however, be emphasized that seismometers need to operate simultaneously over a period of years to provide constraints on the interior structure. Given the high priority that COMPLEX places on the range of objectives that seismic experiments could accomplish, the committee strongly recommends that NASA support passive seismic experiments.

REFERENCES

1. W.M. Folkner, C.F. Yoder, D.N. Yuan, E.M. Standish, and R.A. Preston, "Interior Structure and Seasonal Mass Redistribution of Mars from Radio Tracking of Mars Pathfinder," *Science* 278: 1749–1752, 1997.
2. M.H. Acuña, J.E.P. Connerney, N.F. Ness, R.P. Lin, D. Mitchell, C.W. Carlson, J. McFadden, K.A. Anderson, H. Reme, C. Mazelle, D. Vignes, P. Wasilewski, and P. Cloutier, "Global Distribution of Crustal Magnetization Discovered by the Mars Global Surveyor MAG/ER Experiment," *Science* 284: 790–793, 1999.
3. M.T. Zuber, S.C. Solomon, R.J. Phillips, D.E. Smith, G.L. Tyler, O. Aharonson, G. Balmino, W.B. Banerdt, J.W. Head, C.L. Johnson, F.G. Lemoine, P.J. McGovern, G.A. Neumann, D.D. Rowlands, and S.J. Zhong, "Internal Structure and Early Thermal Evolution of Mars from Mars Global Surveyor Topography and Gravity," *Science* 287: 1788–1793, 2000.
4. R.J. Phillips, M.T. Zuber, S.C. Solomon, M.P. Golombek, B.M. Jakosky, W.B. Banerdt, D.E. Smith, R.M.E. Williams, B.M. Hynek, O. Aharonson, and S.A. Hauck, "Ancient Geodynamics and Global-Scale Hydrology on Mars," *Science* 291: 2587–2591, 2001.
5. See, for example, H. Wänke and G. Dreibus, "Chemical Composition and Accretion History of Terrestrial Planets," *Philosophical Transactions of the Royal Society of London* A235: 545–557, 1988.
6. W.M. Folkner, C.F. Yoder, D.N. Yuan, E.M. Standish, and R.A. Preston, "Interior Structure and Seasonal Mass Redistribution of Mars from Radio Tracking of Mars Pathfinder," *Science* 278: 1749–1752, 1997.
7. See, for example, C.M. Bertka, and Y. Fei, "Implications of Mars Pathfinder Data for the Accretion History of the Terrestrial Planets," *Science* 281: 1838–1840, 1998; C.M. Bertka and Y. Fei, "Density Profile of an SNC Model Martian Interior and the Moment-of-Inertia Factor of Mars," *Earth and Planetary Science Letters* 157: 79–88, 1998; F. Shol and T. Spohn, "The Interior Structure of Mars: Implications from SNC Meteorites," *Journal of Geophysical Research* 102: 1613–1635, 1997; and D.H. Johnston and M.N. Toksöz, "Internal Structure and Properties of Mars," *Icarus* 32: 73–84, 1977.
8. H. Moritz, "Fundamental Geodetic Constraints," *Travaux de L'Association Internationale de Géodésie* 25: 411–418, 1976.
9. C.M. Bertka and Y. Fei, "Implications of Mars Pathfinder Data for the Accretion History of the Terrestrial Planets," *Science* 281: 1838–1840, 1998.
10. D.H. Johnston and M.N. Toksöz, "Internal Structure and Properties of Mars," *Icarus* 32: 73–84, 1977.
11. T. McGetchin and J.R. Smythe, "The Mantle of Mars: Some Possible Geological Implications of its High Density," *Icarus* 34: 512–536, 1978.
12. C.M. Bertka and Y. Fei, "Density Profile of an SNC Model Martian Interior and the Moment-of-Inertia Factor of Mars," *Earth and Planetary Science Letters* 157: 79–88, 1998.
13. A.E. Ringwood, *Composition and Petrology of the Earth's Mantle*, McGraw-Hill, New York, 1975.

14. E. Anders, "Chemical Processes in the Early Solar System, as Inferred from Meteorites," *Accounts of Chemical Research* 1: 289–298, 1968.
15. J.W. Morgan and E. Anders, "Chemical Composition of Mars," *Geochimica et Cosmochimica Acta* 43: 1601–1610, 1979.
16. See, for example, S.J. Weidenschilling, "Accretion of Terrestrial Planets. II.," *Icarus* 27: 161–170, 1976.
17. A.E. Ringwood and S.P. Clark, "Internal Constitution of Mars," *Nature* 234: 89–92, 1971.
18. M.H. Acuña, J.E.P. Connerney, N.F. Ness, R.P. Lin, D. Mitchell, C.W. Carlson, J. McFadden, K.A. Anderson, H. Reme, C. Mazelle, D. Vignes, P. Wasilewski, and P. Cloutier, "Global Distribution of Crustal Magnetization Discovered by the Mars Global Surveyor MAG/ER Experiment," *Science* 284: 790–793, 1999.
19. J.E.P. Connerney, M.H. Acuña, P.J. Wasilewski, N.F. Ness, H. Reme, C. Mazelle, D. Vignes, R.P. Lin, D.L. Mitchell, and P.A. Cloutier, "Magnetic Lineations in the Ancient Crust of Mars," *Science* 284: 794–798, 1999.
20. J.E.P. Connerney, M.H. Acuña, P.J. Wasilewski, N.F. Ness, H. Reme, C. Mazelle, D. Vignes, R.P. Lin, D.L. Mitchell, and P.A. Cloutier, "Magnetic Lineations in the Ancient Crust of Mars," *Science* 284: 794–798, 1999.
21. F. Nimmo, "Dike Intrusion as a Possible Cause of Linear Martian Magnetic Anomalies," *Geology* 28: 391–394, 2000.
22. M.H. Acuña, J.E.P. Connerney, N.F. Ness, R.P. Lin, D. Mitchell, C.W. Carlson, J. McFadden, K.A. Anderson, H. Reme, C. Mazelle, D. Vignes, P. Wasilewski, and P. Cloutier, "Global Distribution of Crustal Magnetization Discovered by the Mars Global Surveyor MAG/ER Experiment," *Science* 284: 790–793, 1999.
23. See, for example, D.E. Smith, M.T. Zuber, S.C. Solomon, R.J. Phillips, J.W. Head, J.B. Garvin, W.B. Banerdt, D.O. Muhleman, G.H. Pettengill, G.A. Neumann, F.G. Lemoine, J.B. Abshire, O. Aharonson, C.D. Brown, S.A. Hauck, A.B. Ivanov, P.J. McGovern, H.J. Zwally, and T.C. Duxbury, "The Global Topography of Mars and Implications for Surface Evolution," *Science* 284: 1495–1503, 1999; and H.V. Frey, S.E. Sakimoto, and J. Roark, "The MOLA Topographic Signature at the Crustal Dichotomy Boundary Zone on Mars," *Geophysical Research Letters* 25: 4409–4412, 1998.
24. D.E. Wilhelms and S.W. Squyers, "The Martian Hemispheric Dichotomy May Be Due to a Giant Impact," *Nature* 309: 138–140, 1984.
25. H.V. Frey and R.A. Schultz, "Large Impact Basins and the Mega-Impact Origin for the Crustal Dichotomy on Mars," *Geophysical Research Letters* 15(3): 229–232, 1988.
26. N. Sleep, "Martian Plate Tectonics," *Journal of Geophysical Research* 99: 5639–5656, 1994.
27. F. Nimmo, and D. Stevenson, "Influence of Early Plate Tectonics on the Thermal Evolution and Magnetic Fields of Mars," *Journal of Geophysical Research* 105: 11969–11980, 2000.
28. R.J. Phillips, M.T. Zuber, S.C. Solomon, M.P. Golombek, B.M. Jakosky, W.B. Banerdt, D.E. Smith, R.M.E. Williams, B.M. Hynek, O. Aharonson, and S.A. Hauck, "Ancient Geodynamics and Global-Scale Hydrology on Mars," *Science* 291: 2587–2591, 2001.
29. M.T. Zuber, S.C. Solomon, R.J. Phillips, D.E. Smith, G.L. Tyler, O. Aharonson, G. Balmino, W.B. Banerdt, J.W. Head, C.L. Johnson, F.G. Lemoine, P.J. McGovern, G.A. Neumann, D.D. Rowlands, and S.J. Zhong, "Internal Structure and Early Thermal Evolution of Mars from Mars Global Surveyor Topography and Gravity," *Science* 287: 1788–1793, 2000.
30. M.T. Zuber, S.C. Solomon, R.J. Phillips, D.E. Smith, G.L. Tyler, O. Aharonson, G. Balmino, W.B. Banerdt, J.W. Head, C.L. Johnson, F.G. Lemoine, P.J. McGovern, G.A. Neumann, D.D. Rowlands, and S.J. Zhong, "Internal Structure and Early Thermal Evolution of Mars from Mars Global Surveyor Topography and Gravity," *Science* 287: 1788–1793, 2000.
31. See, for example, F. Shol, and T. Spohn, "The Interior Structure of Mars: Implications from SNC Meteorites," *Journal of Geophysical Research* 102: 1613–1635, 1997.
32. M.T. Zuber, S.C. Solomon, R.J. Phillips, D.E. Smith, G.L. Tyler, O. Aharonson, G. Balmino, W.B. Banerdt, J.W. Head, C.L. Johnson, F.G. Lemoine, P.J. McGovern, G.A. Neumann, D.D. Rowlands, and S.J. Zhong, "Internal Structure and Early Thermal Evolution of Mars from Mars Global Surveyor Topography and Gravity," *Science* 287: 1788–1793, 2000.
33. L.E. Nyquist, L.E. Borg, and C.-Y. Shih, "The Shergottite Age Paradox and the Relative Probabilities for Martian Meteorites of Differing Ages," *Journal of Geophysical Research* 103: 31445–31455, 1998.
34. R.J. Willemann and D.L. Turcotte, "The Role of Lithospheric Stress in the Support of the Tharsis Rise," *Journal of Geophysical Research* 87: 9793–9801, 1982.
35. R.J. Phillips, M.T. Zuber, S.C. Solomon, M.P. Golombek, B.M. Jakosky, W.B. Banerdt, D.E. Smith, R.M.E. Williams, B.M. Hynek, O. Aharonson, and S.A. Hauck, "Ancient Geodynamics and Global-Scale Hydrology on Mars," *Science* 291: 2587–2591, 2001.
36. W.S. Kiefer, B.G. Bills, and R.S. Nerem, "An Inversion of Gravity and Topography for Mantle and Crustal Structure on Mars," *Journal of Geophysical Research* 101: 9239–9252, 1996.
37. R.J. Phillips, M.T. Zuber, S.C. Solomon, M.P. Golombek, B.M. Jakosky, W.B. Banerdt, D.E. Smith, R.M.E. Williams, B.M. Hynek, O. Aharonson, and S.A. Hauck, "Ancient Geodynamics and Global-Scale Hydrology on Mars," *Science* 291: 2587–2591, 2001.

38. R.J. Phillips, M.T. Zuber, S.C. Solomon, M.P. Golombek, B.M. Jakosky, W.B. Banerdt, D.E. Smith, R.M.E. Williams, B.M. Hynek, O. Aharonson, and S.A. Hauck, "Ancient Geodynamics and Global-Scale Hydrology on Mars," *Science* 291: 2587–2591, 2001.
39. F. Nimmo and D. Stevenson, "Influence of Early Plate Tectonics on the Thermal Evolution and Magnetic Fields of Mars," *Journal of Geophysical Research* 105: 11969–11980, 2000.
40. S.C. Solomon, D.L. Anderson, W.B. Banerdt, R.G. Butler, P.M. Davis, F.K. Duennebier, Y. Nakamura, E.A. Okal, and R.J. Phillips, *Scientific Rationale and Requirements for a Global Seismic Network on Mars*, LPI Technical Report 91-02, Lunar and Planetary Institute, Houston, Texas, 1990.
41. D.L. Anderson, W.F. Miller, G.V. Latham, Y. Nakamura, N.M. Toksoz, A.M. Dainty, F.K. Dunnebier, A.R. Lazarewietz, R.L. Kovach, and T.C.D. Knight, "Seismology on Mars," *Journal of Geophysical Research* 82: 4524–4546, 1977.
42. S.C. Solomon, D.L. Anderson, W.B. Banerdt, R.G. Butler, P.M. Davis, F.K. Duennebier, Y. Nakamura, E.A. Okal, and R.J. Phillips, *Scientific Rationale and Requirements for a Global Seismic Network on Mars*, LPI Technical Report 91-02, Lunar and Planetary Institute, Houston, Texas, 1990.

3

Geochemistry and Petrology

PRESENT STATE OF KNOWLEDGE

Most of what is known about the composition of Mars comes from three types of measurements:

1. In situ analysis of the rocks and regolith[a] by landers;
2. Obital observations carried out by emission and reflectance spectroscopy; and
3. Studies of meteorites that are inferred to have come from Mars.

These sources provide some zero-order compositional data on elemental concentrations in rocks and regolith at a few discrete sites on the planet; limited basic characterization of the global distribution of rock types; and some very detailed knowledge of rocks presumed to come from Mars (the meteorites), which have no geologic context. Much of the first-order information needed to understand the origin and evolution of Mars is still missing.

Elemental Compositions

This subsection discusses only remote and in situ analyses. Far more is known about the compositions of the few dozen meteorites presumed to have come from Mars, but using this information to better understand Mars is difficult because of the lack of geologic context. (The subsection "Knowledge Based on Martian Meteorites" in this chapter focuses on the subject.)

The two Viking landers in the 1970s each carried an x-ray fluorescence spectrometer. These instruments returned data on the major-element composition of the regolith fines at their respective sites in the northern hemisphere.[1] Mars Pathfinder (in 1996) had an Alpha-Proton-X-ray Spectrometer (APXS), which analyzed rock and regolith samples via alpha backscatter, proton-induced x-ray emission, and x-ray fluorescence.[2] The Pathfinder lander returned data on both rocks and regolith fines. The latter were similar in composition to the fines at the two Viking sites (see Figure 3.1a), but the former were richer in silica than the rocks at the Viking sites, and were described as being andesitic in composition (see Figure 3.1b). The normative[b] mineralogies of the samples

[a]The pulverized rock debris that covers most of Mars's surface.

[b]A *norm* is the thermodynamically stable set of minerals that can be made from the proportions of chemical elements found in the bulk chemical analysis of a particular rock.

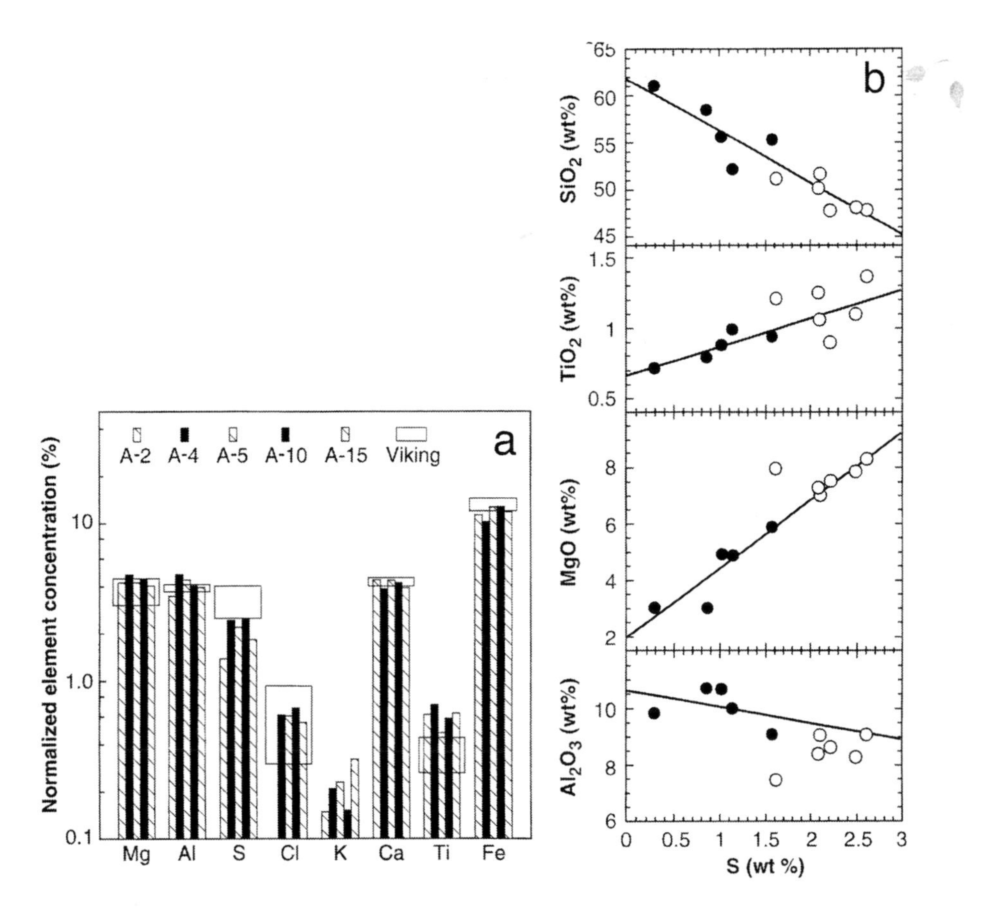

FIGURE 3.1 Comparison of the chemical composition of martian soils as measured on Viking and Pathfinder missions. (a) Comparison of five measured Pathfinder soils (bars) and Viking soil data (boxes). (b) Linear regression lines for several elements versus S in Pathfinder rocks (filled symbols). Soils (open symbols) were not included in the regression but plot on the regression line, indicating that the zero-S values represent a soil-free rock composition. This composition is higher in SiO_2 than are the soils and is indicative of an andesitic rock type. SOURCE: Reprinted with permission from R.T. Rieder, T. Economou, H. Wänke, A. Turkevich, J. Crisp, J. Brückner, G. Dreibus, and H.Y. McSween, Jr., "The Chemical Composition of Martian Soil and Rocks Returned by the Mobile Alpha Proton X-ray Spectrometer: Preliminary Results from the X-ray Mode," *Science* 278: 1771–1774, 1997. Copyright 1997 by the American Association for the Advancement of Science.

are dominated by feldspars, orthopyroxenes, and quartz, but because only a chemical analysis was performed, the mineralogy remains uncertain. The rocks may not even be igneous.

The rocks at the Pathfinder landing site are much more silica-rich[c] than any of the martian meteorites. The regolith fines at all three sites are much higher than the rocks in Mg and Fe, and are similar to the meteorites, but with the addition of S and Cl. The fines at the Pathfinder landing site require the addition of Fe and Mg in some form, perhaps from basalts such as the shergottite meteorites, for their compositions to be explained if they are derived from the Pathfinder rocks. The enhanced abundances of S and Cl in the fines suggest that there may have been mobility of water-soluble compounds in the near surface (see Chapter 5).

[c] SiO_2 content, which increases with the degree of igneous fractionation that preceded formation of a rock, is a key parameter in the description and classification of rocks.

Mineralogical Compositions and Rock Types

As noted above, the Viking and Pathfinder missions provided no direct determination of the mineralogy of the rocks at any of the martian landing sites. However, Mars Global Surveyor carried a Thermal Emission Spectrometer (TES), which returned many spectra of the martian surface.[3] The spectra over most of the planet seem to fall into two broad categories, one indicative of basalt and the other indicative of andesite, consistent with the Viking and Pathfinder analyses. The andesitic composition is more prominent in the northern lowlands, and the basaltic in the older southern highlands. Analysis of the silica-rich rocks at the Pathfinder site indicates that their composition is similar to the average composition of Earth's crust[4] except for their higher Fe content, which may simply reflect the higher Fe content of the martian mantle compared with Earth's.[5]

The TES spectra conclusively identified one major mineral, a coarsely crystalline, gray hematite (α-Fe_2O_3), which is concentrated in a small region near the equator, Sinus Meridiani.[6] This mineral is significant in that it may be diagnostic of large-scale water interactions. TES is sensitive to a wide range of silicate minerals, and analysis of the TES data suggests a significant component of feldspar, as well as high-calcium pyroxene, in both the basaltic and andesitic lithologies.[7] There also appears to be a large component of high-silica glass in the andesite regions, and definitive observations of olivine have been made in the basaltic region of Syrtis Major.[8,9]

Visible/near-infrared observations of Mars have revealed properties of the iron minerals at the surface. Bands at 1 and 2 μm observed in telescopic reflectance spectra indicate the presence of pyroxene in basaltic rocks,[10,11] and orbital observations indicate the presence of both low- and high-calcium pyroxene, consistent with mineralogy of the basaltic SNC meteorites.[12] The red color of Mars is due to the presence of ferric oxide minerals. Bell and colleagues demonstrated that crystalline hematite is present in abundances of up to 4 percent in some bright regions,[13] while much of the ferric mineralogy is best explained by nanophase hematite.[14] These forms of hematite are distinct from the gray hematite observed by TES for a restricted region of the surface. The overall spectral properties across the visible/near-infrared can be fit with palagonite,[d] which suggests that much of the surface is composed of poorly crystallized alteration products. Although carbonates, sulfates, or hydrated minerals have not been definitively observed by any spacecraft measurements to date, the observations have been hampered by poor spatial resolution and/or poor spectral resolution.

Knowledge Based on Martian Meteorites

Members of the SNC category of meteorites described in Chapter 1, comprising the shergottites, nakhlites, and chassignites, plus the unique meteorite ALH84001, are thought to have come from Mars. A variety of data has been used to make this inference, but the strongest case has been made by comparison between the composition of the martian atmosphere as measured by the Viking missions, and the composition of gas trapped by shock in a glass component of the shergottite EETA79001.[15] Because of the importance of these meteorites, they have been studied in great detail. Relating all that is known from these meteorites to what they can tell us about Mars is beyond the scope of this report.[16]

Five different rock types are known in the SNC collection. They are basalts and lherzolites (shergottites), clinopyroxenites (nakhlites), a dunite (Chassigny), and an orthopyroxenite (ALH84001). Most appear to be igneous cumulates. Interestingly, none of these rocks matches the composition of the andesites found at the Mars Pathfinder landing site. The basaltic meteorites have a high Fe/Mg ratio and low Al content relative to terrestrial basalts, and all of the SNC meteorites have oxide minerals consistent with formation under highly oxidizing conditions.

The compositions of these meteorites have been used by two different groups to estimate the bulk composition of Mars.[17,18,19] Though the models disagree on the extent of fractionation among the refractory elements, both models agree that the composition of the martian mantle is similar to that of Earth but with some important differences: Mars is richer in Fe^{3+} and in abundances of moderately volatile elements. Estimates of the size and composition of the core have been made by Treiman and colleagues by accounting for the calculated abundances

[d]A poorly defined alteration product of basalt.

of siderophile and chalcophile elements in the mantle.[20] These estimates are within the ranges allowed by the mean density and moment of inertia of Mars (see Chapter 2).

McSween has discussed various estimates of the major-element compositions of the melts from which the SNC meteorites formed.[21] There are significant difficulties in estimating the melt composition from rocks that are igneous cumulates, because they contain cumulus minerals in abundance greater than what would crystallize from the melt. Nevertheless, it seems clear that all of the magmas have low Al contents, which yield small amounts of plagioclase that forms late in the crystallization sequence of minerals from the melt, and they are all high in Fe.

Trace-element compositions of parent magmas are a little easier to calculate than are major-element compositions, as one can measure the trace-element composition of a cumulus mineral grain and calculate the composition of the coexisting liquid using mineral/liquid partition coefficients. Though the absolute values of the partition coefficients for a given mineral vary with temperature and melt composition, ratios of partition coefficients are not nearly as variable. Thus, the pattern of trace-element abundances can be calculated with some confidence, even if the absolute abundances cannot. McSween has reviewed these calculations.[22] All shergottite parent magmas, as well as the parent of the unique ALH84001, are depleted in light rare-earth elements (LREEs), but the parent magmas of Nakhla and Chassigny are LREE-enriched. Longhi has estimated the abundances of incompatible elements other than the rare-earth elements and found that, like the rare-earth elements, the patterns of these elements in the parent magmas of Shergotty and Nakhla were complementary; that is, the elements most enriched in the Nakhla liquid are most depleted in the Shergotty liquid.[23] This observation suggests that both parent magmas could have come from the same source at different times.

Knowledge of the composition of the SNC parent magmas is important, as it allows one to estimate the composition of the source regions of the magmas. Jones[24] and Longhi,[25] using estimated magma compositions and constraints provided by radiogenic isotopic composition, concluded that the sources for the SNC magmas had time-integrated LREE depletions significantly greater than those of terrestrial magmas. Small degrees of melting of this source could produce the parent magmas of the nakhlites and Chassigny. The shergottite parent magmas could then have formed later from this same source, allowing for the removal of the nakhlite parent liquid, thus accounting for the complementarity described above. An interesting aspect of this model is that all the SNC meteorites could have been derived from a similar source.

NEAR-TERM OPPORTUNITIES

Mars Odyssey

The Mars Odyssey spacecraft was launched in April 2001 and entered orbit about Mars 6 months later. It carries the Thermal Emission Imaging System (THEMIS) and the Gamma-Ray Spectrometer (GRS), both of which will address issues related to geochemistry and petrology.

THEMIS images the martian surface using multispectral, thermal-infrared spectral bands. These images have a spatial resolution of 100 m/pixel, which permits features to be observed at much smaller scale than with the 3-km resolution of TES on Mars Global Surveyor. The high spatial resolution comes at the expense of spectral resolution, which is poorer than that of TES. The diminished spectral resolution may be offset by an ability to discover rock outcrops that are too small to be seen by TES; such outcrops might be of a pure rock type and not a mixture of different rocks or rocks and dust. The poorer spectral resolution may not allow for the determination of the mineralogy, but it should permit mapping of the rock types and other spectral features found by TES. In addition, perhaps other minerals associated with hydrous alteration may be found in areas that are much smaller than the hematite region found by TES.

The GRS will determine the elemental abundances of the surface materials to an accuracy of about 10 percent with a spatial resolution that varies for different elements but is no better than the order of several hundred kilometers.[26] This is not a fine enough spatial scale to sample many of the smaller geologic features, but it will permit global and regional geologic features to be analyzed. It should, for example, be able to determine the chemical differences between the ancient southern highlands and the younger northern lowlands. The chemical composition of the volcanic provinces can also be determined, to see how they differ from the surrounding highlands. Interpretation of the data may be complicated by the ever-present dust component, but the analyses of pure regolith fines by both Viking and Mars Pathfinder may allow a correction for this component on the basis of an element that is enriched in the fines, for example, chlorine. It may be possible to calculate the normative mineralogy from the chemical composition, as was done for the rocks analyzed by Pathfinder; but in an environment where significant amounts of weathering may have occurred, such a calculated mineralogy could be very misleading.

Mars Exploration Rovers—2003

The Mars Exploration Rovers (MERs), scheduled for launch in late May and early June 2003, will each deliver a lander with a rover that is much larger and more scientifically capable than was the Mars Pathfinder rover. The twin MERs will be delivered to sites thought to be good candidates for a future Mars sample-return mission. They will contain a panoramic camera (Pancam); an Alpha Particle X-ray Spectrometer (APXS), similar to that on Pathfinder, for chemical analyses; a Mini-TES for mineralogical analyses; and a microscopic imager for textural analyses to support the mineralogical and elemental analyses. They will also employ a coring device to permit the study of rocks free from weathering products, and a Mössbauer spectrometer for the study of specific iron-bearing phases. This investigation will increase understanding by obtaining data from new sites; additionally, the data will be more comprehensive than that of earlier missions.

RECOMMENDED SCIENTIFIC PRIORITIES

Earlier COMPLEX recommendations of strategies for the exploration of Mars placed strong emphasis on chemical and petrological studies of the planet. Studies of this type provide key information for understanding the processes responsible for the evolution of Mars and its potential as an abode for life. For example, the committee's 1978 report, *Strategy for Exploration of the Inner Planets: 1977–1987*, gives the highest priority to the following objective: "to establish the chemical, mineralogical, and petrological character of different components of the surface material, representative of the known diversity of the planet" (Appendix B: [1.3]).[27] This includes the following (Appendix B: [1.4]):

1. Gross chemical analysis (all principal chemical elements with a sensitivity of 0.1 percent by atom and an accuracy of at least 0.5 atom percent for the major constituents).
2. Identification of the principal mineral phases present (i.e., those making up at least 90 percent of the material in soils and rocks).
3. Establishment of a classification of rocks (igneous, sedimentary, and metamorphic) and fines that define martian petrogenetic processes.
4. Characterization of the state of oxidation, particularly of the fine material and rock surfaces.
5. Characterization of the content of volatiles or volatile-producing species (H_2O, SO_3, CO_2, NO_2).
6. Determination of the selected minor and trace-element contents. (a) Primordial radionuclides: K with a sensitivity of at least 0.05 percent; U and Th with a sensitivity of at least 1 ppm. (b) Selected minor and trace elements (e.g., C, N, F, P, S, Cl, Ti, Ni, As, and rare-earth elements Bi, Cu, Rb, Sr).
7. Measurement of physical properties (magnetic, and in the case of fines, density and size distribution, and rheological properties).

In addition, in its 1990 report *The Search for Life's Origins*, the Committee on Planetary Biology and Chemical Evolution recommended similar measurements (Appendix B: [2.1]) but added "sedimentological and paleontological studies" and called out specific types of sites "where there is evidence of hydrologic activity in any early clement epoch. . . ."[28] In subsequent reports, COMPLEX added other objectives to its strategy for the exploration of Mars and made specific recommendations for developments to aid in the implementation of the strategy, but it has not changed the level of priority assigned in the original 1978 strategy. For example, in the 1995 NASA publication *An Exobiological Strategy for Mars Exploration* a recommendation was made (Appendix B: [5.2]) for "a sequence of landed missions, beginning with development of a geochemically oriented payload capable of regional chemical and mineralogical analyses, oxidant identification, and volatile element detection."[29] The fact that the original COMPLEX priority has not changed is a testament both to how well conceived the strategy was and to how little Mars has been studied until recently.

ASSESSMENT OF PRIORITIES IN THE MARS EXPLORATION PROGRAM

The Mars Exploration Program has already addressed some of the scientific priorities referred to in the preceding section. Mars Pathfinder's APXS investigation, as outlined above, has satisfied the objective for a gross chemical analysis, but only of one rock type at one locality. Mars Global Surveyor's TES instrument has returned global spectral thermal emission data, from which a mineralogic assessment of low-albedo, basaltic, and basaltic-andesite terrains has been derived.[30,31] However, it has been more problematic to determine the modal mineralogy of other regions, probably because of the confounding effects of the poorly crystalline martian dust, and thus it seems unlikely that detailed quantitative mineralogical data of the type recommended by COMPLEX will come out of that investigation. The Mars Odyssey mission (see Table A.1 in Appendix A of this report) has delivered another thermal emission instrument to Mars—THEMIS—with far better spatial, but poorer spectral, resolution. It is too early to say with certainty now, but it is unlikely that this instrument will enable determination of the mineralogy of the surface to the extent recommended by COMPLEX. The gamma-ray spectrometer will return elemental analyses of global-scale regions, but its data will not satisfy the need to sample the diversity of rock types.

The two MERs planned for launch in 2003 should be able to satisfy the need for a gross chemical analysis, just as the Mars Pathfinder APXS did, and it can be hoped that they will find rock types different from those analyzed previously. The Mössbauer spectrometer, Pancam, and Mini-TES instruments will provide insight into the mineralogy of the different rocks and surfaces analyzed, but the detailed petrographic nature of the samples may be hidden by surface alteration and coatings. The Mars Reconnaissance Orbiter (MRO) to be launched in 2005 will look for signs of water, with very high spatial resolution, and the expected high-spatial-resolution visible/near-infrared spectrometer will provide new information relevant to mineralogy and petrology. Nevertheless, the spatial coverage will not be global, and the wavelength range is not sensitive to the full range of minerals expected in the crust and on the surface. As with the synergism between TES and THEMIS, the OMEGA experiment on the European Space Agency's Mars Express, scheduled for launch in 2003, will provide global coverage at a typical spatial resolution of 1 km/pixel, the opposite of the MRO high-resolution spectrometer with its smaller degree of coverage.

Plans exist for a long-range long-duration advanced rover mission—the Mars Science Laboratory (MSL)[e] — in 2007[f] to pave the way for sample return (see Table A.1). The claim is made that such a mission will help develop technologies that should improve the ability to find and collect high-quality samples for Earth return. NASA is wise to take the time necessary for development of the technologies to improve the ability to select interesting sites for sample-return missions; however, there is little justification on scientific grounds for further reconnaissance, following the missions planned through 2005, before the return of the first samples, in a series of sample-return missions. Based on past recommendations of COMPLEX, which are reemphasized here, it is clear that no major strides toward satisfying COMPLEX's highest-priority science objectives in the area of geochemistry and petrology will be made until samples are returned.

---

[e]Also referred to as the Mars Smart Lander, the Mobile Science Laboratory, and by a variety of other names.

[f]Following the completion of this study, NASA announced that it was delaying the launch of MSL until 2009 to allow time to develop an advanced, radioisotope power system for this mission.

REFERENCES

1. P. Toulmin III, A.K. Baird, B.C. Clark, K. Keil, H.J. Rose, R.P. Christian, H.P. Evans, and W.C. Kelliher, "Geochemical and Mineralogical Interpretation of the Viking Inorganic Chemical Results," *Journal of Geophysical Research* 82: 4625–4633, 1977.
2. R.T. Rieder, T. Economou, H. Wänke, A. Turkevich, J. Crisp, J. Brückner, G. Dreibus, and H.Y. McSween, Jr., "The Chemical Composition of Martian Soil and Rocks Returned by the Mobile Alpha Proton X-ray Spectrometer: Preliminary Results from the X-ray Mode," *Science* 278: 1771–1774, 1997.
3. P.R. Christensen, "Introduction to the Special Section: Mars Global Surveyor Thermal Emission Spectrometer," *Journal of Geophysical Research, Planets* 105: 9507, 2000, and 13 subsequent papers on pp. 9509–9739.
4. R.T. Rieder, T. Economou, H. Wänke, A. Turkevich, J. Crisp, J. Brückner, G. Dreibus, and H.Y. McSween, Jr., "The Chemical Composition of Martian Soil and Rocks Returned by the Mobile Alpha Proton X-ray Spectrometer: Preliminary Results from the X-ray Mode," *Science* 278: 1771–1774, 1997.
5. H. Wänke and G. Driebus, "Chemical Composition and Accretional History of Terrestrial Planets," *Philosophical Transactions of the Royal Society of London* A235: 545–557, 1988.
6. P.R. Christensen, J.L. Bandfield, R.N. Clark, K.S. Edgett, V.E. Hamilton, T. Hoefen, H.H. Kieffer, R.O. Kuzmin, M.D. Lane, M.C. Malin, R.V. Morris, J.C. Pearl, R. Pearson, T.L. Roush, S.W. Ruff, and M.D. Smith, "Detection of Crystalline Hematite Mineralization on Mars by the Thermal Emission Spectrometer: Evidence for Near-Surface Water," *Journal of Geophysical Research* 105: 9623–9642, 2000.
7. J.L. Bandfield, V.E. Hamilton, and P.R. Christensen, "A Global View of Martian Surface Compositions from MGS-TES," *Science* 287: 1626–1630, 2000.
8. V.E. Hamilton, R.V. Morris, and P.R. Christensen, "Determining the Composition of Martian Dust and Soils Using MGS TES: Mid-infrared Emission Spectra of Variable-Composition Palagonites," Lunar and Planetary Science Conference 32, Abstract #2123, 2001.
9. T.M. Hoefen and R.N. Clark, "Compositional Variability of Martian Olivines Using Mars Global Surveyor Thermal Emission Spectra," Lunar and Planetary Science Conference 32, Abstract #2049, 2001.
10. T.B. McCord and J.B. Adams, "Spectral Reflectivity of Mars," *Science* 163: 1058–1060, 1969.
11. R.B. Singer and H.Y. McSween, Jr., "The Igneous Crust of Mars: Compositional Evidence from Remote Sensing and the SNC Meteorites," pp. 709–736 in *Resources of Near-Earth Space*, J.S. Lewis, M.S. Matthews, and M.L. Guerrieri (eds.), University of Arizona Press, Tucson, 1993.
12. J.F. Mustard and J.M. Sunshine, "Seeing Through the Dust: Martian Crustal Heterogeneity and Links to the SNC Meteorites," *Science* 267: 1623–1626, 1995.
13. J.F. Bell, T.B. McCord, and P.D. Owensby, "Observational Evidence of Crystalline Iron Oxides on Mars," *Journal of Geophysical Research* 95: 14447–14461, 1990.
14. R.V. Morris, D.G. Agresti, H.F. Laver, J.A. Newcomb, T.D. Shelfer, and A.V. Murali, "Evidence for Pigmentary Hematite on Mars Based on Optical, Magnetic, and Mössbauer Studies of Superparamagnetic (Nanocrystalline Hematite)," *Journal of Geophysical Research* 94: 2760–2778, 1989.
15. For a discussion see, for example, R.O. Pepin, "On the Origin and Evolution of Terrestrial Planet Atmospheres and Meteoritic Volatiles," *Icarus* 92: 2–79, 1991.
16. For a comprehensive review of what martian meteorites can tell us about Mars see, for example, H.Y. McSween, "What We Have Learned About Mars from SNC Meteorites," *Meteoritics* 29: 757–779, 1994.
17. G. Dreibus and H. Wänke, "Mars, a Volatile-Rich Planet," (abstract), *Meteoritics* 20: 367–381, 1985.
18. G. Dreibus and H. Wänke, "Volatiles on Earth and Mars: A Comparison," *Icarus* 71: 221–245, 1987.
19. A.H. Treiman, M.J. Drake, M.-J. Janssens, R. Wolf, and M. Ebihara, "Core Formation in the Earth and Shergottite Parent Body (SPB): Chemical Evidence from Basalts," *Geochimica et Cosmochimica Acta* 50: 1071–1091, 1986.
20. A.H. Treiman, J.H. Jones, and M.J. Drake, "Core Formation in the Shergottite Parent Body and Comparison with Earth," *Journal of Geophysical Research* 92 (supplement): E627–E632, 1987.
21. H.Y. McSween, "What We Have Learned About Mars from SNC Meteorites," *Meteoritics* 29: 757–779, 1994.
22. H.Y. McSween, "What We Have Learned About Mars from SNC Meteorites," *Meteoritics* 29: 757–779, 1994.
23. J. Longhi, "Complex Magmatic Processes on Mars: Inferences from the SNC Meteorites," pp. 695–709 in *Proceedings of Lunar and Planetary Science, Volume 21*, G. Rhyder and V.L. Sharpton (eds.), Lunar and Planetary Institute, Houston, Texas, 1991.
24. J.H. Jones, "Isotopic Relationships Among the Shergottites, the Nakhlites, and Chassigny," pp. 465–474 in *Proceedings of the 19th Lunar and Planetary Science Conference*, G. Rhyder and V.L. Sharpton (eds.), Cambridge University Press, Cambridge, England, 1989.
25. J. Longhi, "Complex Magmatic Processes on Mars: Inferences from the SNC Meteorites," pp. 695–709 in *Proceedings of Lunar and Planetary Science, Volume 21*, G. Rhyder and V.L. Sharpton (eds.), Lunar and Planetary Institute, Houston, Texas, 1991.

26. W.V. Boynton, J.I., Trombka, W.C. Feldman, J.R. Arnold, P.A.J. Englert, A.E. Metzger, R.C. Reedy, S.W. Squyres, H. Wänke, S.H. Bailey, J. Brückner, J.L. Callas, D.M. Drake, P. Duke, L.G. Evans, E.L. Haines, F.C. McCloskey, H. Mills, C. Shinohara, and R. Starr, "Science Applications of the Mars Observer Gamma-Ray Spectrometer," *Journal of Geophysical Research* 97: 7681–7698, 1992.
27. Space Studies Board, National Research Council, *Strategy for Exploration of the Inner Planets: 1977-1987*, National Academy of Sciences, Washington, D.C., 1978.
28. Space Studies Board, National Research Council, *The Search for Life's Origins: Progress and Future Directions in Planetary Biology and Chemical Evolution*, National Academy Press, Washington, D.C., 1990.
29. Exobiology Program Office, NASA, *An Exobiological Strategy for Mars Exploration*, J. Kerridge (ed.), NASA SP-530, NASA, Washington, D.C., 1995.
30. P.R. Christensen, J.L. Bandfield, M.D. Smith, V.E. Hamilton, and R.N. Clark, "Identification of a Basaltic Component on the Martian Surface from Thermal Emission Spectrometer Data," *Journal of Geophysical Research* 105: 9609–9622, 2000.
31. J.L. Bandfield, V.E. Hamilton, and P.R. Christensen, "A Global View of Martian Surface Compositions from MGS-TES," *Science* 287: 1626–1630, 2000.

4

Stratigraphy and Chronology

PRESENT STATE OF KNOWLEDGE

Relevance of Stratigraphy and Chronology to Understanding Mars

The long and complex geological history of Mars, notably the history of its water, can be unraveled by understanding the relative and absolute ages of the planet's geological units, which have been produced or deposited by the various geological processes that have operated throughout the planet's history. Moreover, understanding the geology of a landing site, and therefore, that of samples examined in situ or returned to Earth, requires placing such samples in their appropriate time-stratigraphic geological contexts. Thus, understanding the stratigraphy of Mars is of high priority, as is the ability to date surface units. The absolute ages of surface units will remain necessarily uncertain until samples from known surface locations have been dated in situ and/or on Earth.

Geologic Units and the Stratigraphic Column

The Mariner 9, Viking orbiter, and Mars Global Surveyor spacecraft have provided a wealth of data from which the types and abundances of Mars's geological units can be surveyed, their relative stratigraphic ages derived, and their absolute ages estimated. The planet's various geological units are distinguished and characterized on the basis of their morphologic, topographic, and spectral properties. Their relative ages are determined through examination of their crosscutting and superpositional relationships and the number of superposed impact craters.[1,2] Analyses of this sort have allowed confident derivation of a stratigraphic column for Mars and corresponding chronological ordering of the major geological events in the planet's history (see Figure 4.1).

The major geological units of Mars, their stratigraphic positions, and their locations and extents are reviewed by Tanaka and colleagues.[3] The major divisions thus recognized are the following:

1. Highland rocks, notably those of the Hellas, Argyre, and Isidis impact basins and the southern high plains;
2. Lowland rocks, notably the northern plains and degraded rocks along the highland/lowland boundary;
3. Volcanic and tectonic regions, including highland volcanic paterae, the Tharsis and Elysium volcanic regions, and Valles Marineris;
4. Channel systems, including those originating near Valles Marineris; and
5. Polar regions, including ice and associated layered deposits.

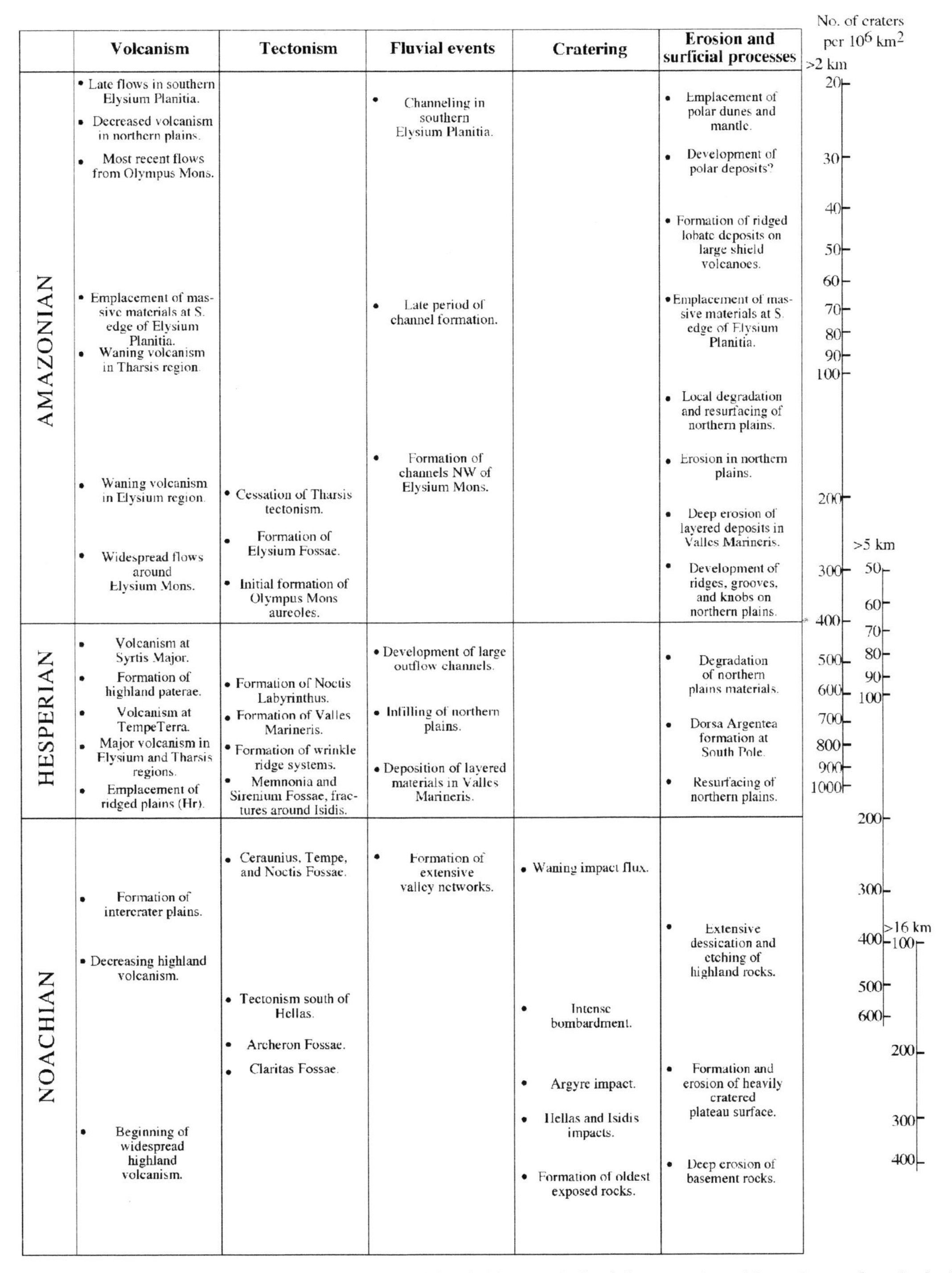

	Volcanism	Tectonism	Fluvial events	Cratering	Erosion and surficial processes
AMAZONIAN	• Late flows in southern Elysium Planitia. • Decreased volcanism in northern plains. • Most recent flows from Olympus Mons. • Emplacement of massive materials at S. edge of Elysium Planitia. • Waning volcanism in Tharsis region. • Waning volcanism in Elysium region. • Widespread flows around Elysium Mons.	• Cessation of Tharsis tectonism. • Formation of Elysium Fossae. • Initial formation of Olympus Mons aureoles.	• Channeling in southern Elysium Planitia. • Late period of channel formation. • Formation of channels NW of Elysium Mons.		• Emplacement of polar dunes and mantle. • Development of polar deposits? • Formation of ridged lobate deposits on large shield volcanoes. • Emplacement of massive materials at S. edge of Elysium Planitia. • Local degradation and resurfacing of northern plains. • Erosion in northern plains. • Deep erosion of layered deposits in Valles Marineris. • Development of ridges, grooves, and knobs on northern plains.
HESPERIAN	• Volcanism at Syrtis Major. • Formation of highland paterae. • Volcanism at TempeTerra. • Major volcanism in Elysium and Tharsis regions. • Emplacement of ridged plains (Hr).	• Formation of Noctis Labyrinthus. • Formation of Valles Marineris. • Formation of wrinkle ridge systems. • Memnonia and Sirenium Fossae, fractures around Isidis.	• Development of large outflow channels. • Infilling of northern plains. • Deposition of layered materials in Valles Marineris.		• Degradation of northern plains materials. • Dorsa Argentea formation at South Pole. • Resurfacing of northern plains.
NOACHIAN	• Formation of intercrater plains. • Decreasing highland volcanism. • Beginning of widespread highland volcanism.	• Ceraunius, Tempe, and Noctis Fossae. • Tectonism south of Hellas. • Archeron Fossae. • Claritas Fossae.	• Formation of extensive valley networks.	• Waning impact flux. • Intense bombardment. • Argyre impact. • Hellas and Isidis impacts. • Formation of oldest exposed rocks.	• Extensive dessication and etching of highland rocks. • Formation and erosion of heavily cratered plateau surface. • Deep erosion of basement rocks.

FIGURE 4.1 Relative ages of major geological events in Mars's history, derived from stratigraphic analyses of geological units and corresponding crater densities (right: expressed as number of craters greater than the indicated diameter per 10^6 km^2 of surface area), grouped here by geological process (SOURCE: J.W. Head III, R. Greeley, M.P. Golombek, W.K. Hartmann, E. Hauber, R. Jaumann, P. Masson, G. Neukum, L.E. Nyquist, and M.H. Carr, "Geological Processes and Evolution," *Space Science Reviews* 96: 263–292, 2001). A correspondence between ***absolute*** ages and crater densities has not been established with confidence. The figure is subject to amendment; it shows Tharsis volcanism to have occurred in the Hesperian epoch, but recent Mars Global Surveyor data have indicated that the Tharsis complex of volcanoes was formed in the upper Noachian epoch.

The planet's geological units are assigned to three major time-stratigraphic systems.[4] The oldest is the Noachian System, named for the ancient rugged materials of Noachis Terra in the southern highlands. Most of the southern highland terrain consists of Noachian-age materials. Rocks of the Hesperian System overlie Noachian units and are characterized by the ridged plains materials of the northern lowlands. Much of the northern lowlands consists of Hesperian age units. The most recent system is the Amazonian, represented by the plains and volcanic materials of Amazonis Planitia. Volcanic materials of the Elysium and Tharsis Montes volcanic regions are Amazonian in age. These three time-stratigraphic systems correspond to three epochs, major time periods having durations and absolute ages that have been estimated from model crater production curves (see the next subsection, "Cratering Chronology").

Data from the Mars Global Surveyor (MGS) spacecraft provide rich insight into martian stratigraphic relationships and the timing of major geological events. Many specific stratigraphic issues have been addressed and new questions have arisen, notably regarding the presence, timing, and extent of liquid water on the surface.[5,6] Moreover, MGS images have revealed thick layered sequences, within Valles Marineris and surrounding troughs and elsewhere across the cratered highlands, which may be volcanic and/or sedimentary in origin.[7,8] These recent observations reveal the rich, active, and wet geological history of early Mars.

Cratering Chronology

The absolute ages of Mars's geological events, and thus the time history of the planet's evolution, will be fully understood only when the relative chronology derived from stratigraphy is tied to an absolute chronology. The density of superposed craters provides a means of estimating absolute chronology, but this technique is dependent on imperfect models of the cratering rate at Mars through time.

Estimates of the impactor flux on Mars through time are based on extrapolation from known lunar fluxes. On the Moon, radiometric dating of samples collected by the Apollo astronauts allows correlation of rocks of known age to impact crater densities observed in images of the sample-collection sites, allowing an absolute chronology to be established for the Moon and extrapolated across the lunar surface.[9] The size-frequency distribution of lunar craters permits derivation of a model curve, known as the "lunar production function," that describes the cratering rate as a function of time.[10] This model curve then must be scaled to the cratering rate as a function of time on Mars.

Adapting the lunar production curve to Mars requires knowledge of the difference in impact rates between the planetary bodies; the nature of crater-forming projectiles; and the effects of differing gravity, impact velocity, and target properties on impact crater formation.[11,12] Modeling of these parameters permits an estimate of the ratio of the lunar to martian production functions, allowing estimation of martian crater ages to within about a factor of two.[13,14]

The factor-of-two uncertainty in age has relatively little effect on interpretation of the absolute age of Noachian terrains, expected to have been originally nearly saturated with craters. Similarly, the factor of two has relatively little effect on interpretation of the age of very young terrains, where a surface with a nominal age of ~10 million years (Myr) is young in any case. However, the factor-of-two uncertainty means that ages of terrains which fall in middle martian history are very poorly constrained.

Model crater ages can constrain the boundaries between Mars's major epochs as defined by its three time-stratigraphic systems. Compilation of various crater age estimates from the literature places the Noachian/Hesperian boundary in the age range of 3.5 billion to 3.8 billion years (Gyr), and the Hesperian/Amazonian boundary within the broad age range of 1.5 to 3.5 Gyr.[15] A recent assessment by Hartmann and Neukum places the Noachian/Hesperian boundary in the nominal age range of 3.5 to 3.7 Gyr, and the Hesperian/Amazonian boundary within the nominal range of 2.9 to 3.3 Gyr.[16] The uncertainty in the martian crater production function implies that late Hesperian through mid-Amazonian ages are the most poorly characterized.

The number of observable martian craters is affected by geological processes of erosion and deposition, with the preservation time of a given crater being dependent on its size and the geological processes that have acted to modify it. Notably, lava flows and eolian activity can partially fill or completely obliterate craters. Eolian activity can exhume a cratered surface that has been protected from cratering for an uncertain amount of time, implying that its crater density would indicate a somewhat younger age than its actual age.[17] While clear evidence for exhumation is found in MGS images, the effect on cratering chronology has not yet been studied in detail. Plainly, understanding the geological context of a region is critical to an accurate estimate of its age based on its crater density.

Absolute Chronology

The SNC meteorites (discussed in Chapters 1 and 3) have contributed useful constraints on surface ages, though their exact provenance on Mars is unknown. Their radiometric ages indicate that some near-surface rocks are as old as 4.5 Gyr and that martian volcanic activity occurred as recently as ~175 Myr ago.[18] These ages are consistent with the interpretation that the heavily cratered highlands date from the earliest history of Mars, and with the interpretation from Mars Global Surveyor images that some very young lava flows have ages of ~10 Myr or less.[19]

Though little is currently known of the absolute chronology of Mars, fruitful future opportunities exist in this area. In situ analysis of rock samples offers a potential means of constraining the absolute chronology of the planet. Analyses of cosmic-ray exposure ages offer the potential for in situ dating of martian rocks in the age range of ~10^5 to 10^7 yr.[20,21,22] This method takes advantage of the fact that a rock within ~1 m of the martian surface has been bombarded by cosmic rays, which through spallation produce nuclei including the noble gases ^{3}He, ^{21}Ne, ^{22}Ne, and ^{38}Ar. Measurement of rock elemental abundances along with these noble-gas abundances can allow an estimate of the production rate of the noble gases and, hence, the length of time the sample has been exposed near the surface. Approximate uniformity of the measured cosmic-ray exposure age of many samples in a region would provide confidence that the crystallization age of the local rock unit has been accurately measured.

K-Ar dating may provide a viable means of dating martian samples with ages >10^6 yr, including those as old as the planet.[23] Using this method in the laboratory to date martian meteorite samples of known radiogenic age, Swindle finds that K-Ar can be used to date samples in situ to an accuracy of ~20 percent.[24] This method assumes that all the ^{40}Ar in a rock sample has been produced by decay of K, and that no ^{40}Ar has escaped the sample over its lifetime. Although Bogard and colleagues argue that K-Ar is not a reliable method for in situ dating on Mars—because atmospheric and cosmogenic ^{40}Ar and ^{36}Ar would confound accurate measurements and calibration,[25] and because ^{40}Ar may be lost from the sample over time—Swindle counters that measurement of ^{36}Ar and ^{38}Ar would allow correction for the atmospheric and cosmogenic contributions of Ar, and that sampling of unshocked and unweathered surface rocks would ensure accurate K-Ar dating for most rock types.[26]

In situ age-dating provides promise for more effectively constraining the present factor-of-two uncertainty in crater-based ages, especially for sites in middle Mars history (late Hesperian through mid-Amazonian). Site(s) dated in situ would better constrain the martian crater production curve, thereby improving estimates of the absolute ages of Mars's other geological units. In order to reduce the uncertainty in rock types sampled and in the interpretation of sample ages, a sampled site or sites should be broad, homogeneous, and geologically well understood. Multiple sites that span middle Mars history would provide the best constraints on the planet's absolute chronology.

Sample return will ensure accurate and precise radiometric age determination, allowing the sample(s) to be subject to intensive laboratory examination. Effective radiometric dating methods could include K-Ar, ^{39}Ar-^{40}Ar, Rb-Sr, Sm-Nd, and U-Th-Pb; by analogy to dating of existing martian meteorites, laboratory precision in age determination would approach 10^7 yr.[27] Accurate understanding of provenance and geological history must be ensured for any returned sample. Only by understanding the geological context can sample ages be generalized to cratering and stratigraphic chronologies and, thus, to the overall history of Mars.

NEAR-TERM OPPORTUNITIES

Orbiter missions provide the opportunity for imaging and spectroscopic studies that can aid stratigraphic and geological analyses and enable improved crater statistics and understanding of geological processes, including erosion and exhumation. The 2001 Mars Odyssey mission carries the Thermal Emission Imaging System (THEMIS) instrument, which will aid in unit characterization (both morphology and composition) through high-resolution imaging and imaging spectroscopy. The Mars Express orbiter's High Resolution Stereo Camera (HRSC) will provide decimeter-scale stereoscopic imaging resolution, greatly aiding stratigraphic studies and refinement of crater-based chronology. NASA's Science Definition Team for the 2005 Mars Reconnaissance Orbiter imaging system has recommended decimeter-scale imaging resolution along with context imaging and spectroscopy, with sufficient swath width (at least 3 km, and preferably 4 to 6 km) to permit local stratigraphy and cratering chronology to be inferred.

Landed missions provide the opportunity to constrain absolute chronology by dating rocks in situ or by collecting and returning samples to Earth for laboratory analyses. The goals of the two 2003 MER missions' rovers do not explicitly address relative or absolute chronology. The specific goals of future lander missions (2007 or 2009 and beyond) have not yet been defined, but could include in situ age measurements and are expected to include sample return. NASA's Mars Scout missions (as yet undefined) also provide potential opportunities for stratigraphic and chronological studies.

RECOMMENDED SCIENTIFIC PRIORITIES

Past recommendations by COMPLEX and other scientific advisory groups regarding strategies for the exploration of Mars have noted the importance of stratigraphy and absolute chronology—they are essential to an understanding of the history and evolution of the planet, notably its water and climate, and to an understanding of the geological context of martian samples.

In its 1978 report *Strategy for Exploration of the Inner Planets 1978–1987*, COMPLEX emphasized that a basic understanding of the times scales of surface materials (Appendix B: [1.2]) and understanding of "the nature and chronology of the major surface forming processes" [1.3] are primary objectives in Mars exploration.[28] Moreover, in its 1990 report *The Search for Life's Origins*, the Committee on Planetary Biology and Chemical Evolution (CPBCE) recommended studies to "[r]econstruct the history of liquid water and its interactions with surface materials on Mars through photogeologic studies, space-based spectral reflectivity measurements, in situ measurements, and analysis of returned samples" [2.2].[29] In discussing the most promising areas from which to return samples, COMPLEX recommended that sample sites be chosen with regard to "the geological context, age, and climatic environment in which the materials were formed" [7.8],[30] and has pointed out the importance of understanding the geological history of martian samples and of the overall planet [9.2].[31] In its 2001 report, NASA's Mars Exploration Payload Assessment Group (MEPAG) also emphasizes the importance of stratigraphic and chronological studies to a broad range of issues relevant to martian geology, climate, and astrobiology [11].[32]

Past reports have recommended both in situ studies and sample return, though their relative emphasis has varied. COMPLEX's 1978 report explicitly recommends that chronological determination should include (1) measurement of "cosmic-ray exposure ages of soil and rock materials for both long and short time scales"; and (2) determination of "crystallization ages of igneous rocks, recrystallization ages of metamorphic rocks, and depositional ages of sedimentary rocks" (Appendix B: [1.5]).[33] The same report notes that sample return "will allow the full range of the most sophisticated analytical techniques to be applied for the study of chronology" [1.19]. The 1990 CPBCE report agrees on the importance of sample return, but emphasizes that even coarse in situ age-dating, which might be accomplished by landed science packages, "can be very valuable in some cases," greatly enhancing our understanding of Mars [4.2].[34] COMPLEX's 1996 letter report "Scientific Assessment of NASA's Mars Sample-Return Mission Options" recommends a focus on sample return to understand Mars as a possible abode of life [6.1].[35] That report notes the overall relevance of in situ measurements for martian exploration and recommends their development [6.2], but does not explicitly refer to in situ age-dating. NASA's 1996 Mars Expeditions Strategy Group report points out that in situ surface studies are essential,[36] but in order to employ appropriately sophisticated and high-precision experiments, "the essential analyses of selected samples must be done in laboratories on Earth" [7.3].

ASSESSMENT OF PRIORITIES IN THE MARS EXPLORATION PROGRAM

The Mars Exploration Program outlined in Appendix A includes the broad goals of understanding the climate and the geological history of the planet within the overarching theme "Follow the water." These goals encompass the objectives of understanding the location and nature of ancient warm and wet environments and of understanding how the planet's climate operated in the more distant past. The Mars Exploration Program explicitly recognizes the importance of understanding the timing of events in Mars history, and emphasizes that the planet may be used as a location from which to provide absolute calibration of the timing of major solar system events.

COMPLEX reiterates past committee recommendations that place high priority on understanding the stratigraphic history and absolute chronology of Mars. Priority should be given to those studies that relate to the evolutionary history of the planet's heat and water and, thereby, to the planet's astrobiological potential.

The planned suite of NASA missions to Mars addresses these important issues in part. Very high resolution imaging, as planned for the Mars Reconnaissance Orbiter and Mars Express missions, could permit relative age-dating of Mars's youngest terrains through analysis of crater distributions at the meter scale, evaluation of the nature of geological boundaries, and better understanding of the active small-scale geological processes that affect the surface. High-resolution imaging also plays a practical role in Mars exploration, being essential to considerations of safety, choice of landing sites, establishment of the context for in situ analyses and returned samples, navigational support for rovers, and so on. The Mars Reconnaissance Orbiter imager and the Mars Express HRSC are expected to accomplish these goals for geological units of interest, and also to provide context imaging and spectroscopy. In general, it is important that very high resolution observations obtain contiguous coverage sufficient to derive meaningful crater statistics, and/or that simultaneous context images be obtained. However, even following these planned missions, portions of the surface of Mars may remain unimaged at sufficient resolution to characterize geological units and determine local stratigraphy and crater-based chronology.

Astronomical observations and dynamical studies of asteroidal and cometary bodies are quite pertinent to improved understanding of the martian crater production function and, thus, its cratering chronology; such studies should be pursued.

Landed missions will allow samples to be dated in situ or to be collected and returned to Earth for much more precise analyses. In situ dating is a developing and as yet unproved technique, but one that shows promise and could be a cost-effective means of constraining absolute chronology. Moreover, analysis of many samples could reduce the error inherent in this measurement technique.

> ***Recommendation.*** COMPLEX recommends that studies of the feasibility of in situ determination of rock ages, by robotic spacecraft, be pursued.

In situ dating could constrain the surface emplacement ages for multiple sites of igneous activity on Mars. This might be achieved through several surface lander and/or rover missions, each of which would examine multiple rock samples within a locale of well-understood geological context and crater density. By constraining the absolute age of several specific geological units, this method can tie the crater age of the planet's geological units to an absolute chronology, revealing the absolute timing of events in martian history. For the purpose of establishing an absolute chronology, sites formed in middle Mars history should have initial priority for age-dating. While the 1978 COMPLEX report *Strategy for Exploration of the Inner Planets* explicitly recommends age determination for soil, igneous rock, and metamorphic rock samples (Appendix B: [1.2, 1.3, 1.5]),[37] this committee concludes that age-dating of igneous rock samples should have clear priority in constraining the emplacement age of a given terrain.

Rock ages determined in situ would be complementary to age-dates acquired by analysis of samples returned to Earth. Sample return will permit precise and accurate age-dating drawing on techniques that cannot practically be used in situ, and will provide significant additional benefits (see Chapter 11 of this report). Overall, the NASA strategy of "Seek, In Situ, Sample" is a sound one with respect to understanding the relative and absolute timing of events in Mars's history.

REFERENCES

1. K.L. Tanaka, D.H. Scott, and R. Greeley, "Global Stratigraphy," pp. 354–382 in *Mars*, H.H. Kieffer, B.M. Jakosky, C.W. Synder, and M.S. Matthews (eds.), University of Arizona Press, Tucson, 1992.
2. J.W. Head, R. Greeley, M.P. Golombek, W.K. Hartmann, E. Hauber, R. Jaumann, P. Masson, G. Neukum, L.E. Nyquist, and M.H. Carr, "Geological Processes and Evolution," *Space Science Reviews* 96: 263–292, 2001.

3. K.L. Tanaka, D.H. Scott, and R. Greeley, "Global Stratigraphy," pp. 354–382 in *Mars*, H.H. Kieffer, B.M. Jakosky, C.W. Synder, and M.S. Matthews (eds.), University of Arizona Press, Tucson, 1992.
4. K.L. Tanaka, D.H. Scott, and R. Greeley, "Global Stratigraphy," pp. 354–382 in *Mars*, H.H. Kieffer, B.M. Jakosky, C.W. Synder, and M.S. Matthews (eds.), University of Arizona Press, Tucson, 1992.
5. J.W. Head, R. Greeley, M.P. Golombek, W.K. Hartmann, E. Hauber, R. Jaumann, P. Masson, G. Neukum, L.E. Nyquist, and M.H. Carr, "Geological Processes and Evolution," *Space Science Reviews* 96: 263–292, 2001.
6. S.C. Solomon, C.L. Johnson, J.W. Head, M.T. Zuber, G.A. Neumann, O. Aharonson, R.J. Phillips, D.E. Smith, H.V. Frey, M.P. Golombek, W.B. Banerdt, M.H. Carr, and B.M. Jakosky, "What Happened When on Mars?: Some Insights into the Timing of Major Events from Mars Global Surveyor Data," Abst. P31A-06, *Eos* 82(20), Spring meeting suppl., 2001.
7. A.S. McEwen, M.C. Malin, M.H. Carr, and W.K. Hartmann, "Voluminous Volcanism on Early Mars Revealed in Valles Marineris," *Nature* 397: 584–586, 1999.
8. M.C. Malin and K.S. Edgett, "Sedimentary Rocks of Early Mars," *Science* 290: 1927–1937, 2001.
9. D. Stöffler and G. Ryder, "Stratigraphy and Isotope Ages of Lunar Geologic Units: Chronological Standard for the Inner Solar System," *Space Science Reviews* 96: 9–54, 2001.
10. G. Neukum, B. Ivanov, and W.K. Hartmann, "Cratering Records in the Inner Solar System in Relation to the Lunar Reference System," *Space Science Reviews* 96: 55–86, 2001.
11. R.G. Strom, S.K. Croft, and N.G. Barlow, "The Martian Impact Cratering Record," pp. 383–423 in *Mars*, H.H. Kieffer, B.M. Jakosky, C.W. Synder, and M.S. Matthews (eds.), University of Arizona Press, Tucson, 1992.
12. B. Ivanov, "Mars/Moon Cratering Rate Ratio Estimate," *Space Science Reviews* 96: 87–104, 2001.
13. B. Ivanov, "Mars/Moon Cratering Rate Ratio Estimate," *Space Science Reviews* 96: 87–104, 2001.
14. W.K. Hartmann and G. Neukum, "Cratering Chronology and the Evolution of Mars," *Space Science Reviews* 96: 165–194, 2001.
15. J.W. Head, R. Greeley, M.P. Golombek, W.K. Hartmann, E. Hauber, R. Jaumann, P. Masson, G. Neukum, L.E. Nyquist, and M.H. Carr, "Geological Processes and Evolution," *Space Science Reviews* 96: 263–292, 2001.
16. W.K. Hartmann and G. Neukum, "Cratering Chronology and the Evolution of Mars," *Space Science Reviews* 96: 165–194, 2001.
17. R. Greeley, R.O. Kuzmin, and R.M. Haberle, "Aeolian Processes and Their Effects on Understanding the Chronology of Mars," *Space Science Reviews* 96: 393–404, 2001.
18. L.E. Nyquist, D.D. Bogard, C.-Y. Shih, A. Greshake, D. Stöffler, and O. Eugster, "Ages and Geologic Histories of Martian Meteorites," *Space Science Reviews* 96: 105–164, 2001.
19. W.K Hartmann and D.C. Berman, "Elysium Planitia Lava Flows: Crater Count Chronology and Geological Implications," *Journal of Geophysical Research* 105: 15011–15025, 2000.
20. See, for example, K. Marti, and T. Graf, "Cosmic-ray Exposure History of Ordinary Chondrites," *Annual Reviews of Earth and Planetary Science* 20: 221–243, 1992.
21. T.D. Swindle, "In Situ Noble-gas Based Chronology on Mars," pp. 294–295 in *Concepts and Approaches for Mars Exploration*, LPI Contribution #1062, 2000.
22. T.D. Swindle, "Applying Noble-gas Geochronology Techniques In Situ on Planets and Asteroids," Eleventh Annual Goldschmidt Conference, Abstract #3718 (CD-ROM), 2001.
23. T.D. Swindle, "In Situ Noble-gas Based Chronology on Mars," pp. 294–295 in *Concepts and Approaches for Mars Exploration*, LPI Contribution #1062, 2000.
24. T.D. Swindle, "Could In Situ Dating Work on Mars?", 31st Lunar and Planetary Science Conference, Abstract #1492 (CD-ROM), 2001.
25. D.D. Bogard, J.L. Birck, O. Eugster, A. Greshake, W.K. Hartmann, G. Neukum, L. Nyquist, M. Ott, G. Ryder, D. Stöffler, and G. Turner, "Letter to the Mars Exploration Program Assessment Group (MEPAG)," Working Group on the Chronology of Mars and the Inner Solar System, International Space Science Institute, Bern, Switzerland, Apr. 15, 2000.
26. T.D. Swindle, "Could In Situ Dating Work on Mars?", 31st Lunar and Planetary Science Conference, Abstract #1492 (CD-ROM), 2001.
27. L.E. Nyquist, D.D. Bogard, C.-Y. Shih, A. Greshake, D. Stöffler, and O. Eugster, "Ages and Geologic Histories of Martian Meteorites," *Space Science Reviews* 96: 105–164, 2001.
28. Space Science Board, National Research Council, *Strategy for Exploration of the Inner Planets 1978–1987*, National Academy of Sciences, Washington, D.C., 1978.
29. Space Studies Board, National Research Council, *The Search for Life's Origins: Progress and Future Directions in Planetary Biology and Chemical Evolution*, National Academy Press, Washington, D.C., 1990.
30. Space Science Board, National Research Council, "Scientific Assessment of NASA's Mars Sample-Return Mission Options," National Research Council, Washington, D.C., 1996.
31. Space Science Board, National Research Council, "Assessment of NASA's Mars Exploration Architecture," National Research Council, Washington, D.C., 1998.

32. NASA, Mars Exploration Payload Assessment Group (MEPAG), "Mars Exploration Program: Scientific Goals, Objectives, Investigations, and Priorities, in Science Planning for Exploring Mars," JPL Publication 01-7, Jet Propulsion Laboratory, Pasadena, Calif., 2001.
33. Space Science Board, National Research Council, *Strategy for Exploration of the Inner Planets 1978–1987*, National Academy of Sciences, Washington, D.C., 1978.
34. Space Studies Board, National Research Council, *The Search for Life's Origins: Progress and Future Directions in Planetary Biology and Chemical Evolution*, National Academy Press, Washington, D.C., 1990.
35. Space Science Board, National Research Council, "Scientific Assessment of NASA's Mars Sample-Return Mission Options," National Research Council, Washington, D.C., 1996.
36. Mars Expeditions Strategy Group, National Aeronautics and Space Administration (NASA), "The Search for Evidence of Life on Mars," 1996, available online at <http://geology.asu.edu/~jfarmer/mccleese.htm>. Also available in National Aeronautics and Space Administration, *Science Planning for Exploring Mars*, JPL Publication 01-7, Jet Propulsion Laboratory, Pasadena, Calif., 2001.
37. Space Science Board, National Research Council, *Strategy for Exploration of the Inner Planets 1978–1987,* National Academy of Sciences, Washington, D.C., 1978.

5

Surface Processes and Geomorphology

PRESENT STATE OF KNOWLEDGE

The surface of Mars is an integrated record of the geological processes that have acted on the planet over its history. The geomorphic landforms provide evidence for constructional events as well as vast erosional episodes in Mars's history. By analogy with geomorphic features on Earth, it has been determined that volcanism, impact cratering, wind, and water have been fundamental drivers of surface modification, and broad constraints have been placed on the relative importance of these geological processes through time. The chemistry and mineralogy of surface materials provide additional constraints not only on the nature of the processes but also on the physical conditions present on or near the surface (e.g., temperature, pH, and humidity).

Water

Evidence that water has been a significant force in shaping the martian surface was revealed by images obtained by the Mariner 9 spacecraft.[1] Our understanding was dramatically expanded by the comprehensive imaging of the planet during the Viking missions. Morphologic features attributed to water can be broadly classified either as formations resulting from running water or as formations resulting from standing bodies of water. Fluvial features range in size from the giant martian outflow channels to valley networks to recently identified small, young channels.[2,3] Morphologic features indicative of standing bodies of water similarly range from putative shoreline features in the northern hemisphere, perhaps resulting from an ocean,[4] to deltaic and intracrater sediments, to finely layered bedding. It is important to note, however, that some of these features are equivocal as evidence for specific processes involving water, or even that water was involved in their formation. The next subsections review the current understanding of these features (see Figure 5.1) and new information derived from the Mars Global Surveyor (MGS) mission.

Giant Outflow Channels

Giant outflow channels—several tens of kilometers across and many hundreds to thousands of kilometers in length—appear to have been cut by enormous floods.[5,6] The channels are mostly Hersperian in age (see Figure 4.1), though some may be as young as Amazonian. The channels commonly start in chaotic terrain, from canyons

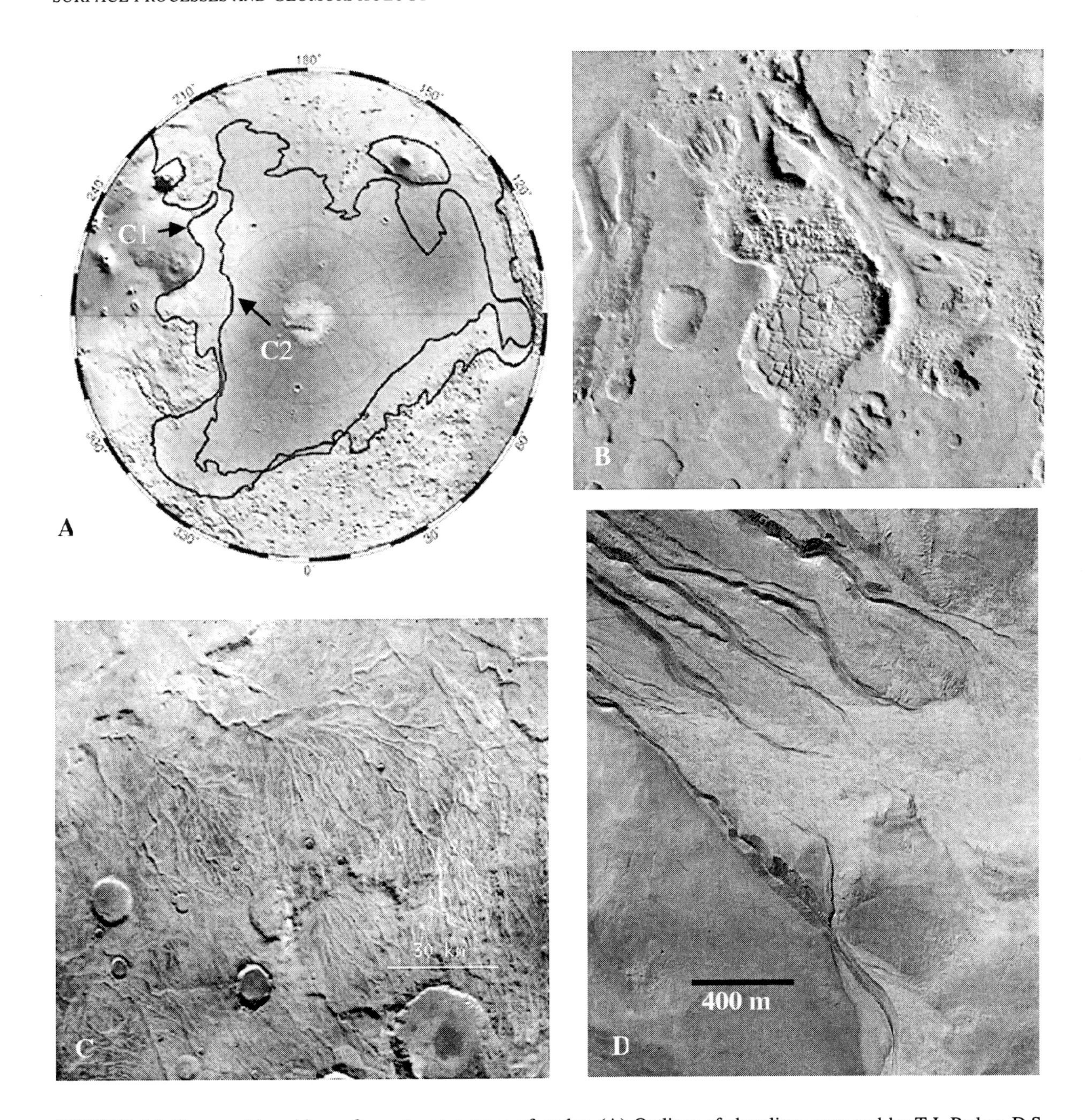

FIGURE 5.1 Geomorphic evidence for water at a range of scales. (A) Outlines of shorelines proposed by T.J. Parker, D.S. Gorcine, R.S. Saunders, D.C. Pieri, and D.M. Schneeberger ("Coastal Geomorphology of the Martian Northern Plains," *Journal of Geophysical Research* 98: 11061–11078, 1993, copyright 1993 by the American Geophysical Union) for a north polar ocean, drawn on a Mars Orbiter Laser Altimeter representation of topography for the northern hemisphere; "C1" refers to Contact 1 and "C2" to Contact 2 (after J.W. Head, H. Hiesinger, M.A. Ivanov, M.A. Kreslavsky, S. Pratt, and B.J. Thomson, "Possible Ancient Oceans on Mars: Evidence from Mars Orbiter Laser Altimeter Data," *Science* 286: 2134–2137, 1999, copyright 1999 by the American Association for the Advancement of Science). Reproduced by permission of AGU and AAAS. (B) Example of outflow channel morphology in Hydaspis Chaos, the source of Tius Valles (Viking Mars Digital Image Model). (C) Valley networks observed by Viking in the Thaumasia region, 42°S, 93°W. (D) Mars Orbiter Camera image M1501466 of youthful channels in a crater near 37°S, 168°W. Images B, C, and D courtesy of NASA/JPL/Malin Space Science Systems.

containing thick, horizontally bedded material interpreted to be sedimentary in origin or from grabenlike depressions. These channels' basic morphology and associated streamlined islands, terraces, and scour strongly support the interpretation that they were formed by massive amounts of flowing fluids, the most likely fluid being water. Lava as an alternative fluid lacks supporting evidence, but CO_2 also has been proposed, along with mechanisms for sequestering CO_2 in the crust and releasing it catastrophically.[7]

The presence of the giant outflow channels and the apparently catastrophic nature of their formation leads to important implications for the outflow channels and to key questions if the channels were formed by water. The channels would require that large volumes of water had been stored in the crust, perhaps within or beneath a thickening cryosphere.[8,9] The water released through disruption of the confining layer or melting of the cryosphere by volcanic processes would have ponded in low points on the planet, which have been postulated to be the northern lowlands, since the channels drain to this point. These occurrences would have had profound impacts on the atmosphere and associated changes in gradational and surface processes.

Evidence for extensive ponding of water in the northern plains was recognized from Viking data (e.g., from sedimentary deposits[10] and shoreline morphology[11]) with the most extreme result of the ponding being a northern ocean. Results from the MGS mission consistent with a northern ocean are that the northern plains inside of Shoreline 2 are extremely smooth,[12] and that the elevation of Shoreline 2 approximates an equipotential surface.[13] However, morphologic features diagnostic of shoreline processes are not immediately evident in the high-resolution Mars Orbiter Camera (MOC) images,[14] though this may be because the observations are not at the relevant scale of the features; additionally, some of the proposed shoreline features identified with Viking images may be tectonic in origin.[15] The origin of the vast, smooth northern plains is nevertheless associated with the catastrophic outflow channels. If the plains are not ocean sediments, they may have been formed from sediments associated with numerous but relatively small outflow events.[16] This subject is critical for unraveling the history of Mars, and fundamental questions remain as to the fate of the outflow channel water and associated sediments, exact timing and duration of the events, and implications for the atmosphere and surface processes.

Valley Networks

Martian valley networks (see Figure 5.1c) superficially resemble the branching patterns of terrestrial fluvial systems, which has led naturally to the interpretation that they were the result of surface runoff under warm, wet climate conditions. Because the valley networks date primarily from the ancient Noachian epoch and predate the outflow channels, they have been cited as evidence that early Mars was warm and relatively wet. However, the channels also exhibit many differences from terrestrial systems, such as lower drainage densities and different morphologies. Analysis of high-resolution MOC data clarifies some of the issues raised by these features.[17] The lack of fine-scale dissection indicates that surface runoff was not a major process in the formation of valley heads. Groundwater sapping is indicated for much of what occurs in the martian valley morphologies.[18] The amount of water required to form the valley networks necessitates some type of recharge to the system, which presents problems discussed by Gulick.[19] The lack of fine dissection implies no surface runoff, and suggests that there was no precipitation or that the infiltration rate was very high. Alternatively, fine dissection features may have been destroyed by surface processes and modification.

While the vast majority of the valley networks are ancient (Noachian), there are also apparently younger valleys of Hesperian age, and an important group of Amazonian-aged networks on the young volcanoes of Alba Patera, Ceraunius Tholus, and Hecates Tholus.[20,21,22] The presence of these channels raises important questions regarding the stability of water in the near surface; recharge of aquifer systems; and the assumed persistence of a cold, dry climate throughout much of Mars's history.[23] The recent discovery of very young gullies on steep slopes poleward of 30° presents an even greater challenge to our notions of the persistence of water in the near surface.[24] The fluids or processes responsible for the formation of these gullies is currently a subject of considerable interest because of the profound implications of modern liquid water on the surface. Alternative fluids, such as CO_2, have been proposed as the agents of erosion,[25] but the plausibility of this model has been sharply criticized.[26] The resolution of this important debate will require definitive compositional or morphological data.

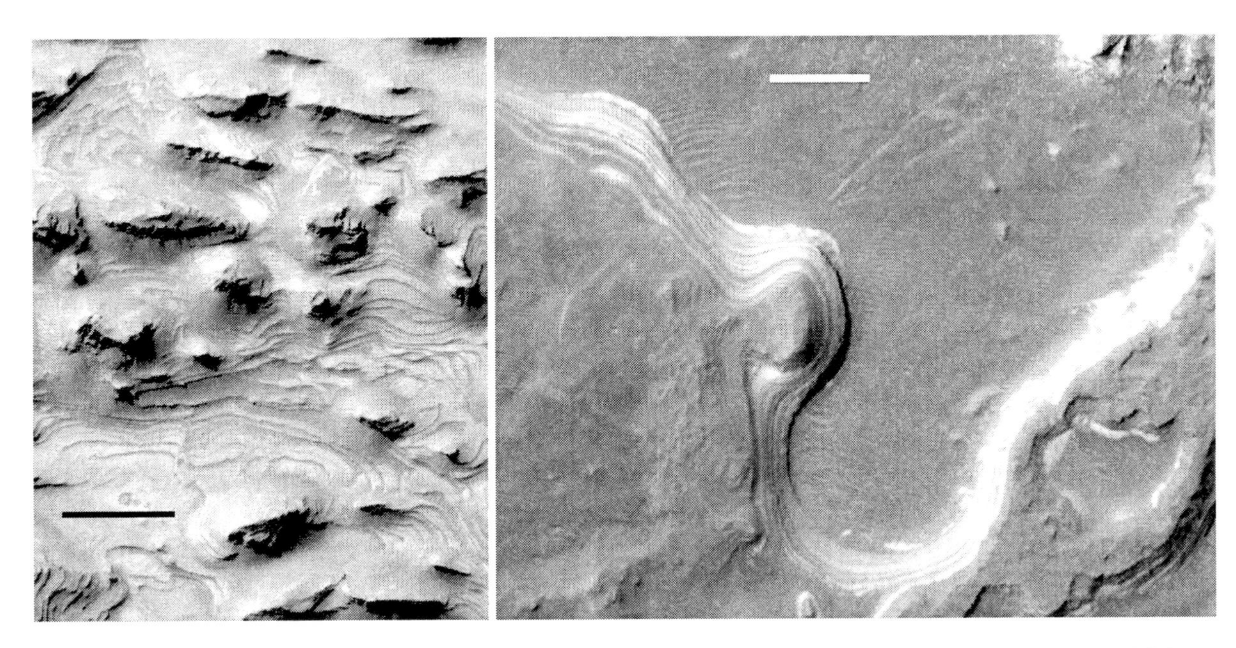

FIGURE 5.2 Fine-scale layering observed with the Mars Orbiter Camera. (Left) Image from Candor Chasma (MOC Image FHA01278). (Right) Image from Holden Crater (MOC Image M0302733). In both images the scale bar is 200 m long. Courtesy of NASA/JPL/Malin Space Science Systems.

Sediments

Sediments deposited in standing bodies of water are high-priority sites for the preservation of fossils and biosignatures.[27] Many of the valley networks terminate in craters, while the outflow channels primarily debouched to the northern plains. Many morphologic features in craters and the northern plains have been interpreted to be either directly (e.g., deltas) or indirectly (e.g., Vastitas Borealas formation in the northern plains) indicative of sedimentary deposits. Thick sequences of layered materials are observed in several of the large canyons, notably the chasmata of Candor and Hebes (see Figure 5.2), which are generally believed to require large standing bodies of water. However, eolian or polar processes may be capable of forming the observed layered materials, and definitive fluvial evidence is still lacking. Data from the Mars Orbiter Laser Altimeter (MOLA) on MGS and high-resolution MOC images have provided a much more refined view of layered materials and their stratigraphic relationships in Valles Marineris and craters.[28] Nevertheless, the morphologic evidence for lacustrine features is equivocal. Carbonates and evaporite minerals have long been predicted to be present on Mars as a natural consequence of lacustrine processes. However, to date there has been no definitive spectral identification of carbonate or other minerals uniquely indicative of evaporite deposits.[29] One detailed study of bright, layered deposits in Pollack Crater (White Rock) showed that the deposits had the same spectral signature as that of bright dust.[30]

Eolian Processes

Wind has been a significant force in shaping the surface of Mars (see Figure 5.3).[31,32] Dunes are ubiquitous features seen across Mars from orbiter to lander resolutions, and so much of the planet exhibits a mantle of fine-grained material that true bedrock exposures are rare. Eolian processes have also been powerful erosive agents, as indicated by specific features such as yardangs[33] and vast regions of etched and eroded terrains.[34] The rapid changes in surface albedo following dust storms attest to the ongoing dynamic modification of the surface by

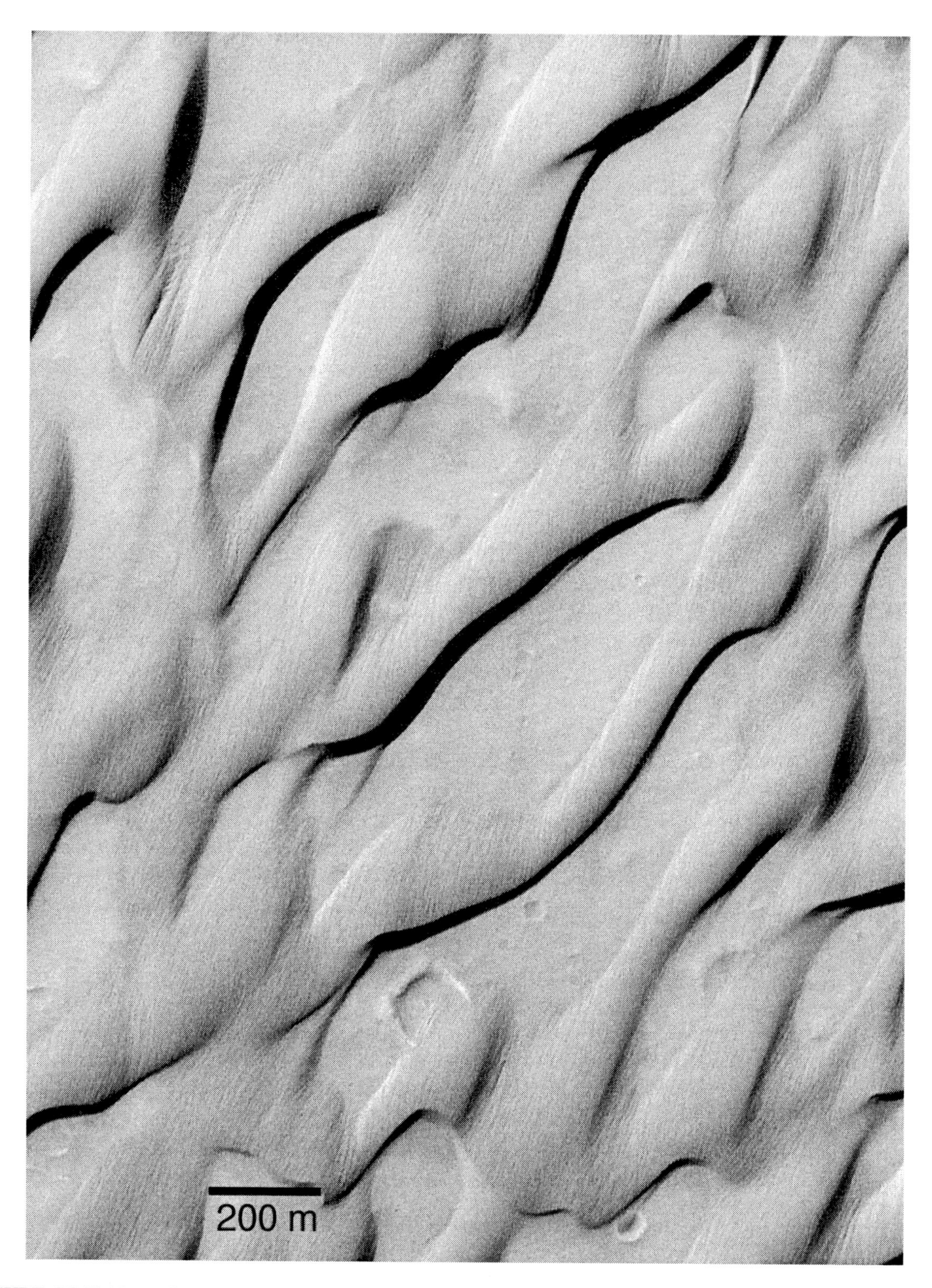

FIGURE 5.3 Evidence for multiple episodes of eolian activity. These sand dunes found in the Herschel Basin of Terra Cimmeria (around 15°S, 228°W) exhibit rough, grooved surfaces, indicating that the dunes are indurated and are undergoing erosion (MOC Image M0003222). Courtesy of NASA/JPL/Malin Space Science Systems.

atmospheric processes. Thus, it is apparent that the surface of Mars has been and continues to be profoundly modified by eolian processes.

Estimates of erosion rates are seemingly at odds with the view from orbit. Bedrock erosion rates at the Viking and Pathfinder sites are too low to explain the dissection of hundreds of meters to kilometers of material in other places on the planet.[35,36] However, the long-term record from the Viking 1 lander attests to the importance of persistent eolian deposition of dust and subsequent removal during storms as a modification process of the surface. This process has the potential to move large amounts of material. In addition to the Viking evidence of infrequent storms, results from Pathfinder quantified the potential importance of dust devils (see Figure 5.4) for disaggregating and transporting material.[37] The large archive of MOC images has yet to be systematically analyzed for eolian processes, but these images are likely to reveal important new information on the characterization and history of eolian processes. A better understanding of the importance of eolian processes through Mars's history will require the following:

1. Thorough characterization of the current atmosphere and its dynamics;
2. Long-term surface observations of the surface and atmosphere at a range of sites;
3. Systematic imaging; and
4. Returned samples.

The surface of Mars provides the palette for understanding eolian processes. However, the surface textures are commonly obscured by the ubiquitous dust cover. Thus, layering and textures that may discriminate among processes are difficult to observe unless exposed in unique situations. Techniques to illuminate the surface beneath this cover, such as imaging radar at a minimum of two wavelengths (e.g., 20 and 70 cm) and two polarizations (e.g., horizontal transmitted, horizontal and vertical received), offer the potential to penetrate the dry dust layer, observe near-surface ice and brines, and reveal the structure and relationships of the near surface. Radar measurements would also contribute substantially to the observations of channel morphology.

Volcanism and Impact Cratering

Most of the surface of Mars bears witness to modification by the processes of volcanism and impact cratering. The myriad features related to volcanism challenge our notions of eruption and surface conditions, and constrain our understanding of Mars's thermal evolution and the contribution of volatiles to the atmosphere from degassing.[38] The style of volcanism changes in space and time across the planet, ranging from the large constructs in the Tharsis region with relatively young surface flows, to the vast Hesparian ridged plains, to the morphologies suggestive of old, explosive volcanism in the central highlands. Our understanding of magma chemistry and absolute chronology is, however, primitive, and it is not yet clear whether the range in volcanic styles represents changes in source regions, changes in near-surface environments, or atmospheric evolution. In addition, there is extensive evidence for magma-volatile interactions in the observed volcanic landforms.[39,40] Topographic and imaging data acquired during the MGS mission, now being assimilated into the prior knowledge base, provide evidence for very ancient as well as very young (<10 million years) volcanism.[41]

Impact cratering has played a central role in developing a chronology for Mars, although this has yet to be calibrated to an absolute age scale (see Chapter 4 in this report). Impacts have also substantially modified the surface and have contributed significantly to the mechanical and chemical weathering of the surface. Because the impact process delivers a substantial heat pulse to the surface and crust, there has been speculation that cratering could be a major process in establishing hydrothermal systems, although no direct evidence of hydrothermal alteration in craters has been observed. Martian impact craters also display a more diverse array of morphologies and preservation states than do the impact craters of any other planet (see Figure 5.5). While these morphologies generally have been thought to be due to impact into volatile-rich surfaces, interaction of ejecta with the atmosphere may also contribute significantly to the observations. New information from the MGS mission is providing a vastly refined view of crater morphology that will likely lead to new insights into the cratering process on Mars. As discussed in Chapter 4, however, a correlation between crater densities and absolute chronology on Mars is still lacking.

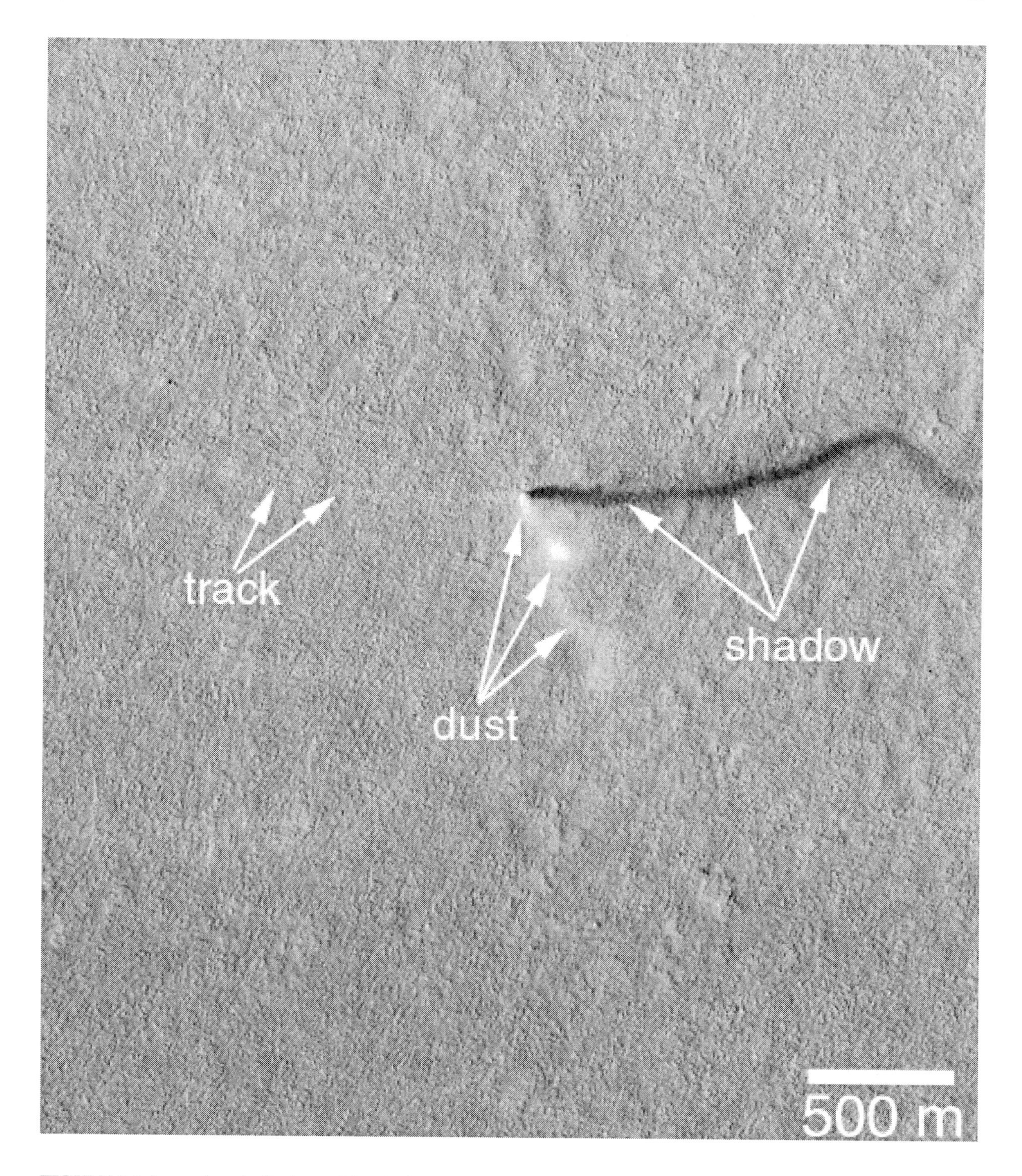

FIGURE 5.4 Large dust devil observed by the Mars Orbiter Camera. The column of dust raised by this dust devil casts a shadow on the martian surface. Based on the Sun direction and shadow length, this dust devil is approximately 1 km in height. A faint track caused by its passage is visible on the surface. SOURCE: MGS MOC Release No. MOC2-281. Courtesy of NASA/JPL/Malin Space Science Systems.

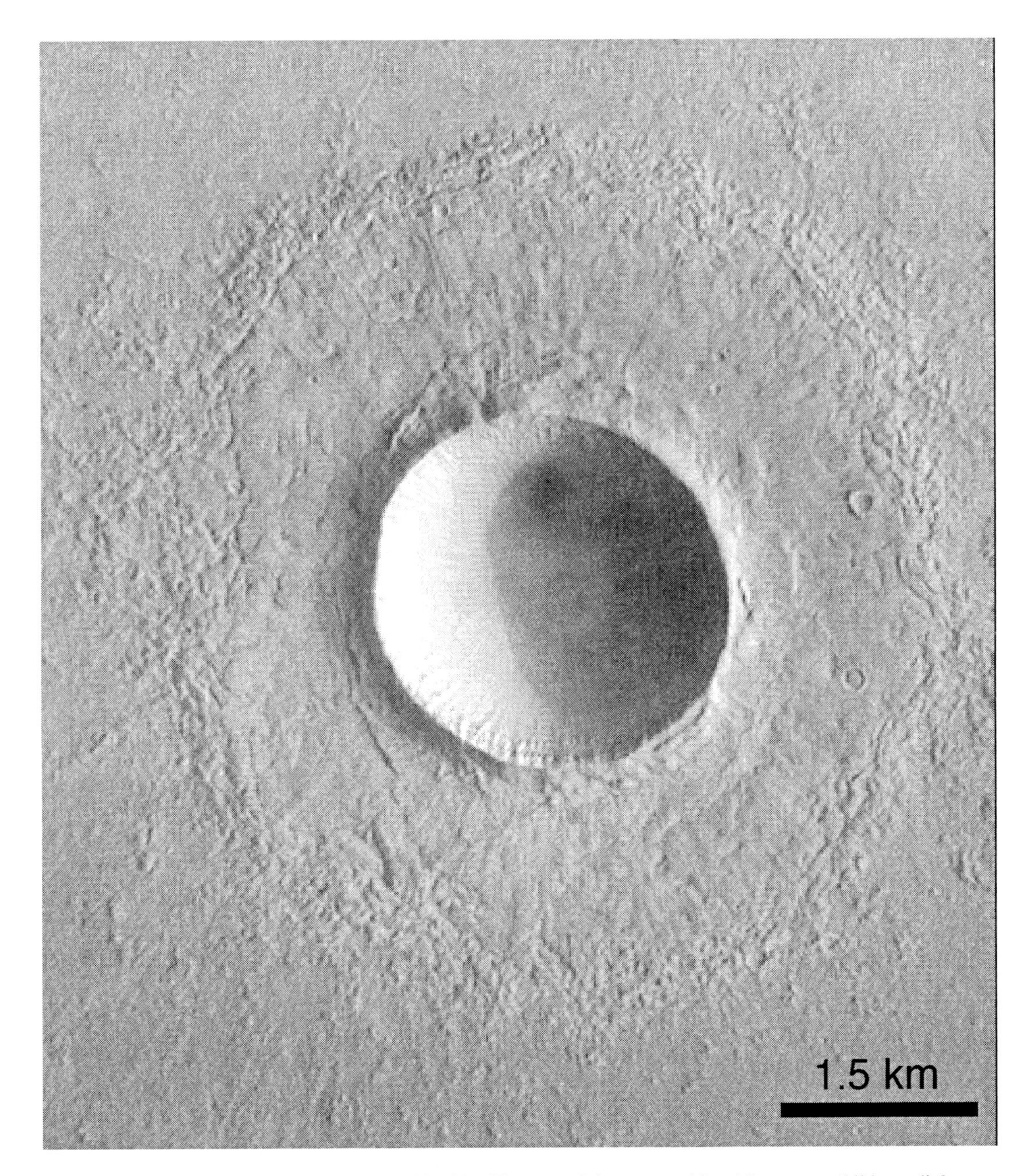

FIGURE 5.5 Impact crater in northern Elysium Planitia. The material excavated by this crater exhibits radial grooves and occupies a fairly thick, well-contained ejecta blanket. Illumination is from the right/upper right. SOURCE: MGS MOC Release No. MOC2-161. Courtesy of NASA/JPL/Malin Space Science Systems.

Physical and Chemical Alteration

The chemical and mineralogical composition of the upper 1 meter to 1 kilometer of Mars contains a record of the history of surface-atmosphere interactions. Much of the surface of Mars (though not all) is clearly oxidized, but the timing, rates, and processes of this oxidation are poorly understood. The chemical compositions of the oxidized mobile materials measured directly by the Viking and Pathfinder missions were remarkably similar, despite the wide geographic separation of the sites. The soils are distinguished by their relatively high Fe and S and low Si and Al contents. Their mineralogy has not been directly measured; the inferred mineralogy consists of poorly crystalline or cryptocrystalline products of basalt alteration. The large amounts of Cl and S in the soils suggest the presence of soluble salts such as sulfates, and the presence of cemented soils or duricrust attests to possible mobility of soluble compounds in the near surface. The general composition of the soils can be explained by hydrolytic weathering of basalt with a significant addition of S and Cl through the atmospheric deposition of volcanic aerosols.[42]

Remotely sensed data provide additional constraints on the surface mineralogy. The general visible/near-infrared spectral properties are well modeled by palagonite, a poorly crystallized product of low-temperature basalt alteration. Only a few definitive mineralogic signatures have been observed. The specific ferric oxide mineralogy and its form are critical information for understanding the chemical and physical pathways of alteration and weathering. The presence of poorly crystallized or nanophase hematite is well supported,[43] and there are indications of hydrated ferric oxides such as ferrihydrite.[44] However, the precise geographic and vertical distributions of these different oxides have not yet been resolved. From Thermal Emission Spectrometer (TES) data, Christensen and colleagues discovered highly localized concentrations of coarsely crystalline gray hematite, very distinct in form and origin from the nanophase hematite.[45] The silicate mineralogy of the soils is currently unknown.

Models for the physical and chemical alteration of Mars span a wide range of possible mechanisms. The presence of an apparently deeply oxidized ancient crust coupled with the apparently unoxidized later volcanic landforms has led to the idea that most of the weathering occurred early, during a warmer, wetter time, and that alteration has been sporadic since then.[46] Estimated rates of weathering under current conditions are essentially negligible.[47] To a large extent, the critical measurements necessary for an understanding of weathering and alteration have not yet been made. Little is known regarding the chemistry and reactivity of the soils (e.g., pH and Eh conditions), or what the exchanges of volatiles are between the atmosphere and the surface.

The stunning images acquired by the MOC instrument on MGS have opened up a new perspective for understanding surface processes and geomorphology,[48,49,50,51,52] and the array of new landforms that has been revealed is stimulating fresh ideas about the evolution of Mars. In addition, while our expectations for the detailed characteristics of many surfaces at these high resolutions have been met (e.g., surfaces that appear bland and highly degraded at moderate resolution are similar at high resolution), others have not been borne out (e.g., surfaces that appear smooth in Viking images may appear rough to MOC, and vice versa). This may be seen as the result of leaping to a high resolution before acquiring adequate knowledge at intermediate resolution (analogous to jumping from use of a hand lens to use of an electron microscope), suggesting the importance of acquiring nested imaging data of appropriately cascading resolutions that will allow for understanding context and scale.

High-Resolution Topography

As the MOLA experiment has demonstrated, highly precise and accurate topography is a critical and fundamental measurement required to understand Mars. While of great vertical precision and accuracy, the MOLA measurements are nevertheless widely spaced and thus not sufficiently dense to contribute to understanding detailed relationships among landforms on the surface. Attaining an understanding of many of the fundamental problems in surface processes and geomorphology will require topographic measurements with spatial resolutions that match the scale of the features under investigation. For example, to estimate the thickness of the intracrater layered materials observed with MOC and to determine their volumes will require measurements of 10-m-or-better spatial resolution and relative altitudes with precision approaching or exceeding MOLA (1 m or better). Stereo imaging (e.g., by the High Resolution Stereo Camera (HRSC) on Mars Express, to be launched in 2003) can provide this information.

NEAR-TERM OPPORTUNITIES

Resolution of the outstanding questions in surface processes and geomorphology will require three basic approaches:

1. Orbital observations of atmospheric processes and of morphology through imaging and topography;
2. Measurement of the spectral properties of the surface at moderate and high spatial resolution, to determine mineralogy; and
3. Landed science investigations to acquire detailed measurements of surface properties, and surface processes, and for the selection of samples for return to Earth.

While continued analysis of archival data from MGS can be expected to contribute significantly to our understanding of surface processes and geomorphology, a number of near-term opportunities will enhance this effort.

Planned Research Opportunities on Future Missions

The suite of orbiter and landed science missions currently in operation, launched, or in development (see Table A.1) provides a number of measurements needed to understand the surface of Mars. The MGS mission has refined the measurement of global and regional topography through MOLA, and novel uses of its capabilities have allowed the measurement of the volatile dynamics of the north polar cap. The MOC database continues to grow, providing additional measurements of morphology at high resolution. The TES instrument will be completing its coverage of the planet at thermal infrared wavelengths and moderate (3 × 6 km/pixel) spatial resolution, which will refine our understanding of surface composition as revealed by this wavelength range and provide information on thermal inertia, atmospheric temperature profiles, and dust loading. The Mars Odyssey orbiter carries the Gamma-Ray Spectrometer (GRS), which will measure the elemental composition of the surface, including H, at coarse spatial resolution, and THEMIS, a multispectral thermal camera that will provide global 100-m/pixel coverage and will distinguish rock types and search for thermal anomalies. In addition, THEMIS has a visible/near-infrared multispectral camera that will provide 10 percent global coverage at 40-m/pixel resolution.

Investigations on the Mars Express orbiter relevant to surface processes include the HRSC and an Infrared Mineralogical Mapping Spectrometer (called OMEGA). HRSC will acquire images in four spectral bands and multiple angles to create global three-dimensional multispectral maps with spatial resolutions between 10 and 30 m and vertical resolution of approximately 20 m, and 2-m/pixel coverage over 1 percent of the planet. OMEGA will cover the wavelength range from 0.5 to 5.2 μm with a spectral resolution sufficient to identify mafic, alteration, and carbonate minerals. It is also sensitive to the degree of hydration of the surface. OMEGA is expected to acquire global coverage at 1 to 4 km/pixel during the lifetime of the mission, with selected regions at higher resolution. The orbiter will also carry a Subsurface Sounding Radar/Altimeter (called MARSIS) to search for indications of water in the top 5 km of the martian crust as well as to investigate near-surface crustal structure. Mars Express will carry a small lander, Beagle 2, equipped with a Mössbauer spectrometer to measure iron mineralogy and oxidation state; x-ray and mass spectrometers to measure elemental composition and carbon isotopes; and instruments to measure atmospheric temperature, pressure, and wind speed and direction.

The instrument payload on the twin Mars Exploration Rovers scheduled for launch in 2003 will include a multispectral camera and a thermal emission spectrometer for imaging and mineralogy of the surface, and a high-resolution camera and Mössbauer and APXS spectrometers to measure the elemental composition and oxidation state of rocks and soils. The instruments on Mars Reconnaissance Orbiter (MRO), planned for launch in 2005, will include a 60-cm/pixel camera; a 50-m/pixel visible/near-infrared imaging spectrometer with spectral capabilities comparable to those of OMEGA; a moderate-resolution (7-m/pixel) context imager; a radar sounder; the Pressure-Modulator Infrared Radiometer (PMIRR) to measure water vapor; and the Mars Color Imager (MARCI), a multichannel ultraviolet and visible camera that will be used to globally and quantitatively map atmospheric O_3, clouds, and hazes.

Opportunities beyond 2005 have been broadly defined by NASA and the international community. The expectation is that in the 2007 launch opportunity NASA will focus on landed science and capable rovers as well as Mars Scout missions, which will be competed in a manner similar to the Discovery program. Exploitation of the 2009 opportunity has not been defined beyond a possible Italian Space Agency-NASA science orbiter. The 2011 opportunity is tentatively allocated to sample return. The mainstay of this approach is to alternate landed and orbital science at each opportunity to take better advantage of discoveries made and lessons learned.

With a successful completion of the orbital observations through the MRO mission, the global reconnaissance mapping of the Mars surface with remotely sensed data will be largely complete. The only major observation that will not be covered is synthetic aperture radar imaging with multiple polarizations and frequencies.

RECOMMENDED SCIENTIFIC PRIORITIES

Understanding past and present distribution of water on Mars remains a critical scientific priority, as identified in reports from COMPLEX and other scientific advisory groups (Appendix B: [1.5, 4.4, 11]). The current and past distribution of water from the surface to depth, processes governing the cycling of water between reservoirs, the past history of these reservoirs, and the record of water as an agent of change for the surface dominate the list of basic scientific priorities for understanding surface processes and geomorphology. Because water is such a pervasive theme, it cuts across all possible measurements, from orbital observations of climate to detailed investigations of returned samples.

Because the surface is the interface between the interior and the atmosphere, it is critical to understand the composition and chemistry of the current atmosphere, as well as its circulation and climate (Appendix B: [1.10, 4.3, 4.5, 4.7]). This is essential for understanding current processes and the potential for the atmosphere to interact with the surface through alteration and exchange of volatiles.

As a chronicle of past processes, the sedimentary record of Mars is a critical priority identified in all previous scientific assessments (Appendix B: [1.5, 4.2, 4.4, 5.2, 11.1, 11.2, 11.4]). Understanding this record will require a global inventory of deposits; characterization of the environments of formation; relative and absolute ages; and detailed measurement of mineralogy, texture, and chemistry.

The timing and duration of major episodes in Mars's history and understanding of their relative importance are required to determine the evolution of the planet (Appendix B: [1.2, 11.2]). Current understanding is based on superposition relationships of surface morphologic units and cratering statistics. To better understand the major formational events and their processes, a precise chronology must be developed for Mars on the basis of isotopic measurements (see Chapter 4). Also, high-resolution topography must be measured with much higher spatial density than MOLA achieved for select regions, to enable understanding of detailed spatial relationships among units, and there must be high-spatial-resolution mineralogy of sites where bedrock is exposed.

The key tool for understanding geomorphic and surface processes is imaging. The current inventory of images ranges from coarse to very fine resolution, and different processes are understood at these different scales and by integrating across scales (Appendix B: [1.1, 11.1]). It is essential to have nested observations of the surface at relevant resolutions (e.g., 1, 10, 100 m) acquired at comparable observing geometries to provide context for very high spatial resolution images of limited coverage and to allow observation of processes that are not revealed at these high resolutions. It is also clear that multispectral imaging is a very important tool in relating information across scales, but this can be done at an intermediate (10- to 20-m) resolution.

The physical record of past climates is expected to be recorded in the mineralogy and composition of near-surface units (see Chapter 3 in this report) (Appendix B: [1.21, 3.2, 11.1, 11.2, 11.3]). This record needs to be measured with high-spatial-resolution spectroscopy, landed science investigations, and returned samples. The current chemistry and mineralogy of the surface should be determined, including the oxidation state, pH, and Eh conditions. The chemistry and mineralogy also provide relevant information for understanding the mechanics of surface processes.

ASSESSMENT OF PRIORITIES IN THE MARS EXPLORATION PROGRAM

The suite of orbital missions and their associated measurements, planned by NASA through the Mars Reconnaissance Orbiter mission and internationally through the Mars Express mission, address many if not most of the scientific priorities relevant to geomorphology and surface processes that can be analyzed from orbit, and identified in previous COMPLEX reports and in recent reports from MEPAG. If all the orbital investigations planned for launch as outlined above are successful and meet the mission objectives, the reconnaissance mapping of Mars will be largely completed, and initial detailed investigations will have begun. The only major missing measurement is multipolarization, multifrequency synthetic aperture radar imaging. Because of constraints on the return of data, high-resolution mapping will not cover much of the planet, but it may be possible to extrapolate high-resolution studies to larger areas through the global mapping efforts.

The distribution of water in the upper crust (see Chapter 6) is expected to be illuminated by sounding radar (e.g., MARSIS on Mars Express). While this technique has not yet been proven to provide unequivocal or even interpretable data, the planned experiment will be a critical demonstration of its capabilities. If the experiment is successful, then a powerful new approach to measuring water in the martian crust will be available for future missions.

Nested imaging is essential for scaling observations, and the MRO mission will be the first to obtain submeter and ~10-m panchromatic imaging along with hyperspectral 20- to 50-m imaging, all acquired at the same observing geometry. However, systematic mapping of the entire surface with color data will be lacking. This was to have been acquired by the MARCI medium-angle camera on the lost Mars Climate Orbiter, but the MARCI instrument on MRO will not have this color capability. Through the extended mission of Mars Express, HRSC will acquire multispectral observations in stereo at 10- to 30-m resolution, and this will provide the first high-resolution topographic measurements (2 m/pixel). It is likely that more observations at this resolution will be required in the future to support the geomorphology and surface process science goals and to support landed science and sample-return missions.

The combination of high-spectral- and high-spatial-resolution imaging by MRO will provide the observations necessary to assess some aspects of the mineralogy of the near surface and its relationship to geomorphology and surface processes. The minerals sensitive to the proposed visible-through-infrared wavelength range of the MRO spectrometer include iron-bearing mafic silicates, sulfates, carbonates, clays, and ferric oxides and oxyhydroxides, but not the full range of silicates (e.g., quartz, feldspar). The sensitivity to detection depends on many factors, such as spectral contrast, and limits identification to the most abundant phases of the most spectrally active minerals. In general, mineral determination from orbit gives global information, but with poor specificity. Spectral studies cannot provide the definitive data that can be obtained by studying samples, but they are important for deciding where to go to get samples.

Landed science investigations, carefully targeted within sites for which the full suite of remotely sensed data are available, will allow extrapolation of detailed field investigations to larger scales. This combination of results will address many, but not all, of the important science goals relevant to the chemical and mineralogical signatures of geomorphic processes and surface evolution. In particular, minerals that do not have distinct spectroscopic signatures or that are present at low abundance will not be detected.

Through the progression of missions to highly capable rovers, the anticipated measurements of mineralogy and chemistry at well-characterized sites will be essential in determining the history of water and the importance of surface-atmosphere interactions. Many important science questions will require these measurements, as well as the study of returned samples. The NASA Mars Exploration Program is directed ultimately toward these objectives.

REFERENCES

1. T.A. Mutch, R.E. Arvidson, J.W. Head, K.L. Jones, and R.S. Saunders, *The Geology of Mars*, Princeton University Press, Princeton, N.J., 1976.
2. M.H. Carr, *Water on Mars*, Oxford University Press, New York, 1996.
3. M.C. Malin and K.S. Edgett, "Evidence for Recent Groundwater Seepage and Surface Runoff on Mars," *Science* 288: 2330–2335, 2000.
4. T.J. Parker, D.S. Gorcine, R.S. Saunders, D.C. Pieri, and D.M. Schneeberger, "Coastal Geomorphology of the Martian Northern Plains," *Journal of Geophysical Research* 98: 11061–11078, 1993.
5. V.R. Baker and D.J. Milton, "Erosion by Catastrophic Floods on Mars and Earth," *Icarus* 23: 27–41, 1974.

6. V.R. Baker, M.H. Carr, V.C. Gulick, C.R. Williams, and M.S. Marley, "Channels and Valley Networks," pp. 493–522 in *Mars*, H.H. Kieffer, B.M. Jakosky, C.W. Synder, and M.S. Matthews (eds.), University of Arizona Press, Tucson, 1992.
7. N. Hoffmann, "White Mars: A New Model for Mars' Surface and Atmosphere Based on CO_2," *Icarus* 146: 326–342, 2000.
8. M.H. Carr, "Formation of Martian Flood Features by Release of Water from Confined Aquifers," *Journal of Geophysical Research* 84: 2995–3007, 1979.
9. S.M. Clifford, "A Model for the Hydrologic and Climatic Behavior of Water on Mars," *Journal of Geophysical Research* 98: 10973–11016, 1993.
10. B.K. Lucchitta, H.M. Ferguson, and C.A. Summers, "Sedimentary Deposits in the Northern Lowland Plains, Mars," *Proceedings of 17th. Lunar and Planetary Science Conference*, printed as a supplement to the *Journal of Geophysical Research* 91: E166–E174, 1986.
11. T.J. Parker, D.S. Gorcine, R.S. Saunders, D.C. Pieri, and D.M. Schneeberger, "Coastal Geomorphology of the Martian Northern Plains, *Journal of Geophysical Research* 98: 11061–11078, 1993.
12. T.J. Parker, D.S. Gorcine, R.S. Saunders, D.C. Pieri, and D.M. Schneeberger, "Coastal Geomorphology of the Martian Northern Plains, *Journal of Geophysical Research* 98: 11061–11078, 1993.
13. J.W. Head, H. Hiesinger, M.A. Ivanov, M.A. Kreslavsky, S. Pratt, and B.J. Thomson, "Possible Ancient Oceans on Mars: Evidence from Mars Orbiter Laser Altimeter Data," *Science* 286: 2134–2137, 1999.
14. M.C. Malin and K.S. Edgett, "Oceans or Seas in the Martian Northern Lowlands: High Resolution Imaging Tests of Proposed Coastlines," *Geophysical Research Letters* 26: 3049–3052, 1999.
15. P. Withers and G.A. Neumann, "Enigmatic Northern Plains of Mars," *Nature* 410: 651, 2001.
16. B.M. Jakosky and R.J. Phillips, "Mars' Volatile and Climate History," *Nature* 412: 237–244, 2001.
17. M.H. Carr and M.C. Malin, "Meter-scale Characteristics of Martian Channels and Valleys," *Icarus* 146: 366–386, 2000.
18. V.C. Gulick, "Origin of the Valley Networks on Mars: A Hydrological Perspective," *Geomorphology* 37: 241–268, 2001.
19. V.C. Gulick, "Origin of the Valley Networks on Mars: A Hydrological Perspective," *Geomorphology* 37: 241–268, 2001.
20. D.H. Scott and J.M. Dohm, "Mars Highland Channels: An Age Reassessment," *Lunar and Planetary Science Conference* 23: 1251–1252, 1992.
21. V.C. Gulick and V.R. Baker, "Origin and Evolution of Valleys on Martian Volcanoes," *Journal of Geophysical Research* 95: 14325–14344, 1990.
22. M.H. Carr, *Water on Mars*, Oxford University Press, New York, 1996.
23. M.T. Mellon and R.J. Phillips, "Recent Gullies on Mars and the Source of Liquid Water," 32nd Lunar and Planetary Science Conference, Abstract #1182 (CD-ROM), 2001.
24. M.C. Malin and K.S. Edgett, "Evidence for Recent Groundwater Seepage and Surface Runoff on Mars," *Science* 288: 2330–2335, 2000.
25. See, for example, D.S. Musselwhite, T.D. Swindle, and J.I. Lunine, "Liquid CO_2 Breakout and the Formation of Recent Small Gullies on Mars," *Geophysical Research Letters* 28: 1283–1285, 2001.
26. S.T. Stewart and F. Nimmo, "Surface Runoff Features on Mars: Testing the Carbon Dioxide Formation Hypothesis," Lunar and Planetary Science Conference 32, Abstract #1780 (CD-ROM), 2001.
27. N.A. Cabrol and E.A. Grin, "Distribution, Classification, and Ages of Martian Impact Crater Lakes," *Icarus* 142: 160–172, 1999.
28. M.C. Malin and K.S. Edgett, "Sedimentary Rocks of Early Mars," *Science* 290: 1927–1937, 2000.
29. P.R. Christensen, J.L. Bandfield, R.N. Clark, K.S. Edgett, V.E. Hamilton, T. Hoefen, H.H. Kieffer, R.O. Kuzmin, M.D. Lane, M.C. Malin, R.V. Morris, J.C. Pearl, R. Pearson, T.L. Roush, S.W. Ruff, and M.D. Smith, "Detection of Crystalline Hematite Mineralization on Mars by the Thermal Emission Spectrometer: Evidence for Near-surface Water," *Journal of Geophysical Research* 105: 9623–9642, 2000.
30. S.W. Ruff, P.R. Christensen, R.N. Clark, H.H. Kieffer, M.C. Malin, J.L. Bandfield, B.M. Jakosky, M.D. Lane, M.T. Mellon, and M.A. Presley, "Mars' 'White Rock' Feature Lacks Evidence of an Aqueous Origin," 31st Lunar and Planetary Science Conference, Abstract #1945 (CD-ROM), 2000.
31. R. Greeley and J.D. Iverson, *Wind as a Geological Process on Earth, Mars, Venus, and Titan*, Cambridge University Press, Cambridge, England, 1985.
32. R. Greeley, N. Lancaster, S. Lee, and P. Thomas, "Martian Aeolian Processes, Sediments and Features," pp. 730–766 in *Mars*, H.H. Kieffer, B.M. Jakosky, C.W. Synder, and M.S. Matthews (eds.), University of Arizona Press, Tucson, 1992.
33. J.F. McCauley, "Mariner 9 Evidence for Wind Erosion in the Equatorial and Mid-latitude Regions of Mars," *Journal of Geophysical Research* 78: 4123–4137, 1973.
34. P.H. Schultz and A.B. Lutz, "Polar Wandering on Mars," *Icarus* 73: 91–141, 1988.

35. R.E Arvidson, J.L. Gooding, and H.J. Moore, "The Martian Surface as Imaged, Sampled, and Analyzed by the Viking Landers," *Reviews of Geophysics and Space Physics* 27: 39–60, 1989.
36. A.W. Ward, L.R. Gaddis, R.L. Kirk, L.A. Soderblom, K.L. Tanaka, M.P. Golombek, T.J. Parker, R. Greeley, and R. Kuzmin-Ruslan, "General Geology and Geomorphology of the Mars Pathfinder Landing Site," *Journal of Geophysical Research* 104: 8555–8571, 1999.
37. S.M. Metzger, J.R. Carr, J.R. Johnson, T.J. Parker, and M.T. Lemmon, "Dust Devil Vortices Seen by the Mars Pathfinder Camera," *Geophysical Research Letters* 26: 2781–2784, 1999.
38. See, for example, P.J. Mouginnis-Mark, L. Wilson, and M.T. Zuber, "The Physical Volcanology of Mars," pp. 424–452 in *Mars*, H.H. Kieffer, B.M. Jakosky, C.W. Synder, and M.S. Matthews (eds.), University of Arizona Press, Tucson, 1992.
39. See, for example, P.J. Mouginnis-Mark, L. Wilson, and M.T. Zuber, "The Physical Volcanology of Mars," pp. 424–452 in *Mars*, H.H. Kieffer, B.M. Jakosky, C.W. Synder, and M.S. Matthews (eds.), University of Arizona Press, Tucson, 1992.
40. See, for example, D.A. Crown and R. Greeley, "Volcanic Geology of Hadriaca Patera and the Eastern Hellas Basin, Mars," *Icarus* 100: 1–25, 1993.
41. W.K. Hartmann and D.C. Berman, "Elysium Planitia Lava Flows: Crater Count Chronology and Geological Implications," *Journal of Geophysical Research* 105: 15011–15025, 2000.
42. A. Banin, B.C. Clark, and H. Wänke, "Surface Chemistry and Mineralogy," pp. 594–625 in *Mars*, H.H. Kieffer, B.M. Jakosky, C.W. Synder, and M.S. Matthews (eds.), University of Arizona Press, Tucson, 1992.
43. J.F. Bell, T.B. McCord, and P.D. Owensby, "Observational Evidence of Crystalline Iron Oxides on Mars," *Journal of Geophysical Research* 95: 14447–14461, 1990.
44. S. Murchie, L. Kirkland, S. Erard, J. Mustard, and M. Robinson, "Near-infrared Spectral Variations of Martian Surface Materials from ISM Imaging Spectrometer Data," *Icarus* 147: 444–472, 2000.
45. P.R. Christensen, J.L. Bandfield, R.N. Clark, K.S. Edgett, V.E. Hamilton, T. Hoefen, H.H. Kieffer, R.O. Kuzmin, M.D. Lane, M.C. Malin, R.V. Morris, J.C. Pearl, R. Pearson, T.L. Roush, S.W. Ruff, and M.D. Smith, "Detection of Crystalline Hematite Mineralization on Mars by the Thermal Emission Spectrometer: Evidence for Near-Surface Water," *Journal of Geophysical Research* 105: 9623–9642, 2000.
46. J.F. Bell, "Iron, Sulfate, Carbonate, and Hydrated Minerals on Mars," pp. 359–380 in *Mineral Spectroscopy: A Tribute to Roger G. Burns*, M.D. Dyar, C. McCammon, and M.W. Schaefer (eds.), Geochemical Society, St. Louis, Missouri, 1996.
47. J.L. Gooding, R.E. Arvidson, and M.Y. Zolotov, "Physical and Chemical Weathering," pp. 626–651 in *Mars*, H.H. Kieffer, B.M. Jakosky, C.W. Synder, and M.S. Matthews (eds.), University of Arizona Press, Tucson, 1992.
48. See, for example, P.C. Thomas, M.C. Malin, M.H. Carr, G.E. Danielson, M.E. Davies, W.K. Hartmann, A.P. Ingersoll, P.B. James, A.S. McEwen, L.A. Soderblom, and J. Veverka, "Bright Dunes on Mars," *Nature* 397: 592–594, 1999.
49. M.C. Malin and K.S. Edgett, "Evidence for Recent Groundwater Seepage and Surface Runoff on Mars," *Science* 288: 2330–2335, 2000.
50. M.C. Malin and K.S. Edgett, "Sedimentary Rocks of Early Mars," *Science* 290: 1927–1937, 2000.
51. M.H. Carr and M.C. Malin, "Meter-Scale Characteristics of Martian Channels and Valleys," *Icarus* 146: 366–386, 2000.
52. W.K. Hartmann and D.C. Berman, "Elysium Planitia Lava Flows: Crater Count Chronology and Geological Implications," *Journal of Geophysical Research* 105: 15011–15025, 2000.

6

Ground Ice, Groundwater, and Hydrology

PRESENT STATE OF KNOWLEDGE

Water on Mars is a major crosscutting theme for martian scientific studies as well as for planned Mars exploration. As on Earth, water exists on Mars in many states and participates in a broad range of important physical, chemical, and possible biological processes. Water has played a key role in the evolution of the martian climate and in the shaping of Mars's geological history. Our present state of knowledge regarding water on Mars can be divided into three categories: (1) present reservoirs, (2) current hydrology, and (3) paleohydrology. A complete understanding of water on Mars will eventually require detailed interdisciplinary study in all three areas.

Present Reservoirs

The question of where water is on Mars today is easy to pose but difficult to answer fully. Direct observations exist of exposed martian water reservoirs, which include water vapor in the atmosphere, water ice in the atmosphere, seasonal water ice deposits at the surface, and permanent water ice deposits at the polar caps.[1,2] Of the four reservoirs, the martian polar caps are by far the most massive. Recent topographic profiles from Mars Global Surveyor's (MGS) Mars Orbiter Laser Altimeter (MOLA) indicate that the mass of water ice contained within Mars's north and south polar caps, assuming a high ice-to-dust ratio, is the equivalent of a global water layer 22 to 33 m thick.[3]

Beyond the water reservoirs that can now be detected on Mars, there is good reason to suspect the presence of hidden water reservoirs whose combined masses should be much greater than those that are currently exposed.[4,5] In Mars's near-surface regolith, one expects water to be adsorbed on soil particles,[6] and there is fragmentary evidence from the Viking Gas Exchange experiment that its mass fraction could be on the order of 1 percent.[7] Water very probably also occurs bound in rocks and regolith materials in the form of hydrated minerals. MGS Thermal Emission Spectrometer (TES) observations demonstrate the presence of small isolated regions rich in crystalline hematite, and analyses of SNC meteorites give evidence for the presence of low-temperature hydrated minerals in crustal rocks and of higher-temperature hydrated minerals deep within the mantle.[8] Geomorphic evidence from Viking and MGS observations indicates that the layered deposits surrounding the north and south polar caps also contain water ice, but its mass fraction is currently not well constrained.[9] One also expects to find near-surface ground ice on Mars, as on Earth;[10,11] models predict that it should be present within the top meters of the surface at latitudes as low as 20 degrees from the equator in favorable locations,[12] but to date no direct measurements exist that constrain its abundance or its geographical distribution. Nevertheless, there is abundant geomorphological evidence for past processes associated with the surface expression of this ice.[13] At kilometer depths, Mars's geothermal gradient should eventually give rise to conditions where temperatures exceed 0° C, and there liquid water will be stable (see Figure 6.1).[14] Although the evidence for the potential existence of deep, liquid water environments on Mars today is compelling, there are as yet no direct measurements of their existence or global distribution. Again, however, there is abundant evidence for past aqueous activity related to this groundwater (see Chapter 7 of this report).[15] In summary, the water observed in the atmosphere and at the surface of Mars is believed to be only "the tip of the iceberg," and significantly larger water reservoirs may lie within reach of future exploration.

Current Hydrology

The behavior of water on Mars is governed by the interaction between its chemical and thermodynamic properties and by the martian environment. Because of Mars's low surface temperatures, the partitioning of water is heavily biased toward its condensed phases, causing the martian atmosphere to be extremely dry and ineffective at transporting large quantities of water on seasonal time scales. Liquid water on Mars is not expected to be stable on Mars today, because temperatures exceed 273 K only at low latitudes during the warmest periods of the day, and any liquid generated would quickly evaporate and be transported by the atmosphere to colder locations where it would then freeze.[16]

The most detailed observations of the behavior of martian water come from measurements of the column abundance of water vapor in the martian atmosphere from the Viking landers' Mars Atmospheric Water Detector instruments,[17,18] and more recently from MGS's TES.[19] These measurements show that atmospheric water vapor abundances reach maximum values of close to 100 precipitable microns at high northern latitudes during the summer season when the north polar water ice cap is exposed. Elsewhere on Mars, column water vapor abundances are generally less than 15 precipitable microns. Atmospheric models based on those developed for the terrestrial atmosphere show that the diurnal and seasonal behavior of water vapor in the atmosphere is affected by a complex range of properties and processes, which include the availability of water sources and water sinks, the thermal structure and dynamics of the atmosphere, and the distribution and behavior of atmospheric aerosols.[20,21] It is currently an open question as to whether the seasonal cycle of water vapor observed today is in net annual equilibrium.

Efforts to understand the present behavior of water on Mars are severely hindered by a lack of information on

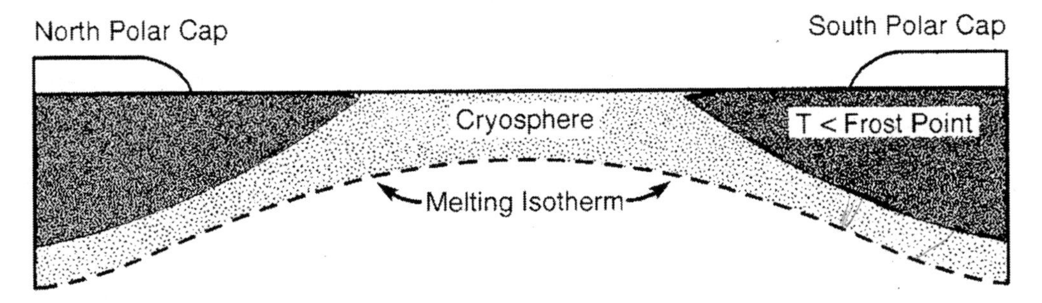

FIGURE 6.1 Diagrammatic pole-to-pole cross section of the martian crust illustrating the theoretical latitudinal variation in the stability of ground ice (shaded zones; cryosphere) and the depth to liquid water stability (melting isotherm). Only in the dark-shaded zones (T < frost point) are temperatures low enough to condense ice from the martian atmosphere. SOURCE: S.M. Clifford, "A Model for the Hydrologic and Climatic Behavior of Water on Mars," *Journal of Geophysical Research* 98: 10973–11016, 1993. Copyright 1993 by the American Geophysical Union. Reproduced by permission of AGU.

the physical and chemical properties of the near-surface environment, as well as by a lack of observations of the three-dimensional behavior of water vapor in the atmosphere. Very complex models for the behavior of water in the martian atmosphere can be constructed. The flux of volatiles in and out of the near-surface sediments is important for these models, but without the acquisition of detailed information regarding the physical and chemical properties of the critical surface/atmosphere interface and the actual fluxes, it will be difficult to provide adequate model constraints. The problem of insufficient information is even more true for efforts to understand the behavior of water beneath the martian surface. At the present time, very little detailed information exists about the adsorptive capacity and diffusivity of the martian regolith,[22] and without this information and an understanding of the behavior of water at the surface/atmosphere interface, it will be difficult to use models to extend our knowledge of the behavior of adsorbed water and near-surface ground ice. With respect to the question of deep liquid water environments, the absence of information regarding Mars's heat-flow rate and the permeability of the deep regolith results in even more severe limitations. In summary, the theoretical modeling tools for understanding the present behavior of water on Mars are in place, but the detailed observations necessary to constrain them are lacking.

Paleohydrology

Some of the most exciting questions concerning Mars deal with the past distribution and behavior of water. Many of these questions are motivated by geomorphic evidence such as runoff channels, outflow channels, and other features that have been interpreted to mean that liquid water may have been present periodically on the surface of Mars in past epochs (see Chapter 7 in this report).[23] MGS Mars Orbiter Camera (MOC) and MOLA observations have provided many types of new evidence for the possible presence of liquid water on Mars. Among these are MOLA observations that refine our understanding of the large channels that once flowed from the southern highlands to the northern lowlands;[24] MOC images showing the presence of widespread ancient layering that is believed to be of sedimentary origin;[25] and small gullies on crater walls that are considered to be evidence for recent erosion by fluids—the most likely considered to be liquid water (see also Chapter 5 in this report).[26] MGS data are also potentially consistent with many other scenarios for the possible history of water on Mars, ranging from that of a relatively dry past[27] all the way to that of a large ocean which completely covered the northern lowlands.[28,29] Most intriguing are indicators of very young (perhaps 1-million to 10-million-year-old) landforms indicative of climatic change associated with the emplacement of near-surface ground ice,[30] the activity of debris-covered glaciers,[31] and contraction-cracked, polygonal terrain,[32] indicating the melting of an active layer over ice-rich permafrost. SNC meteorites contain mineralogic and isotopic evidence for subsurface water-rock interactions.[33] Unfortunately, because of the lack of knowledge regarding the origin of these samples, it is difficult to place this information in a geologic or climatological context.

The multiplicity of current theories regarding the past history of water on Mars point out some of the difficulties in attempting to understand the past purely from an incompletely observed martian geologic record. Another approach is to use the clues provided by the geologic record to guide inquiries into the potential behavior of water on Mars using physical and chemical models. For example, oceans do not just spring out of nowhere: If there were large oceans at some point in Mars's history, there must also have been certain definable environmental conditions that enabled their stability. At present, detailed models for the general atmospheric circulation of Mars exist and are improving[34,35] and match the current observations of the atmosphere by MGS reasonably well.[36] There are areas for improvement in these models, such as accommodation of CO_2 and water-ice clouds.

NEAR-TERM OPPORTUNITIES

The dual failures of Mars Polar Lander (MPL) and Mars Climate Orbiter (MCO) in 1999 resulted in the loss of a significant fraction of our near-term opportunities to study the distribution and behavior of martian water. MPL was to have made the first measurements of the near-surface behavior of water vapor over diurnal and seasonal time scales as well as of the abundance of adsorbed water and water ice in the near-surface regolith, at a very favorable high-latitude location. Simultaneous measurements of atmospheric water vapor from orbit by the Pressure-Modulator Infrared Radiometer (PMIRR) instrument would have made it possible to interpret "ground-truth" water measurements at the Polar Lander's landing site within the context of a global, three-dimensional model of atmospheric structure and circulation over a complete Mars year.

Mars Odyssey's GRS instrument, which was originally flown on the failed Mars Observer in 1993, and its associated neutron spectrometer have the capability of mapping the global abundance of hydrogen in the near-surface regolith, which will provide important constraints on the present distribution of near-surface water. Mars Odyssey also includes the THEMIS instrument, which will search for the presence of hydrated minerals at higher spatial resolution than can be done by the MGS TES.

Looking into the future, the European Space Agency's Mars Express mission, which is scheduled for launch in 2003, will include a long-wavelength radar sounding experiment that will make the first measurements of subsurface electromagnetic properties. These may reveal the presence of subsurface ice layers and liquid water. Mars Express also includes the Beagle 2 lander, whose payload will include environmental-monitoring and soil-analysis instrumentation that may advance our understanding of martian water.

In 2005, NASA plans to launch the Mars Reconnaissance Orbiter (MRO), which will include the third attempted flight of the PMIRR atmospheric sounder, as well as the Mars Color Imager (MARCI) camera (also part of the lost Mars Observer and Mars Climate Orbiter payloads). PMIRR and MARCI will provide important information about water vapor and weather, although the value of the measurements is diminished without the concurrent surface measurements that would have been provided by the MPL. MRO will also include a visible/near-infrared imaging spectrometer that will identify the hydration state of the surface, and will include subsurface sounding radar to probe for water and ice in the crust. Both the Infrared Mineralogical Mapping Spectrometer (OMEGA) of the Mars Express mission and the proposed visible/near-infrared imaging spectrometer of MRO have the capability of identifying hydrated minerals. The European mission NetLander, planned for launch in 2007, will set up a network of seismometers that may reveal the presence of deep, liquid water environments, as well as a network of meteorology stations that may improve understanding of the martian general atmospheric circulation. In summary, despite recent setbacks, there still remain a diverse set of near-term opportunities to further our understanding of the distribution and behavior of water on Mars.

RECOMMENDED SCIENTIFIC PRIORITIES

Determining the distribution, abundances, sources and sinks, and histories of volatile materials has been consistently recommended as a first-order priority for Mars exploration for more than two decades (Appendix B: [1.3, 1.6, 1.9, 1.10, 2, 4, 5.1, 5.2, 7, 10.2, 11.1, 11.2, 11.3]). While the entirety of possible topics that can be addressed here is vast, a number of fundamental observations would fill in significant gaps in understanding (e.g., the three-dimensional distribution of water in the martian crust).

Of the various martian volatile materials, liquid water has been singled out as the highest priority because of the potential for stable liquid water environments to serve as the setting for past and present martian life (Appendix B: [2, 7.1]). The presence of liquid water, its persistence in various environments, and the conditions under which is may exist in near-surface locations are of high importance.

Understanding water on Mars from the perspective of observing its present behavior (Appendix B: [1.3, 1.9, 1.10, 2, 4, 7.2, 11.1, 11.2.1, 11.2.3, 11.2.6, 11.3.3]), and in particular the sources, sinks, and reservoirs of water, is of first-order importance. Such an understanding would provide a solid basis for engaging in the analysis of geological evidence for past behavior (Appendix B: [1.3, 2, 4.4, 5.1, 5.2, 7.2, 7.6, 7.7, 11.1, 11.2.3, 11.2.5, 11.2.6, 11.3.2]), and would thus aid in developing a more integrated view of the history of water on Mars.

ASSESSMENT OF PRIORITIES IN THE MARS EXPLORATION PROGRAM

Searching for water on Mars in its various forms has been a significant component of the Mars program, but the search thus far has not been comprehensive and has lacked balance, in large part because of the failure of three missions. The loss of Mars Observer in 1993 took with it the PMIRR and GRS instruments, which would have provided high-quality information about the present global distribution of water vapor in the atmosphere, and water ice and/or adsorbed water in the near-surface martian regolith. PMIRR was reflown on Mars Climate

Orbiter, but this spacecraft failed in 1999. PMIRR is now scheduled to fly again on the Mars Reconnaissance Orbiter in 2005. The second GRS was launched successfully on Mars Odyssey in 2001 and began science operations in orbit about Mars in 2002. The loss of the Polar Lander in 1999 resulted in the loss of the Mars Volatiles and Climate Surveyor (MVACS) integrated payload, which would have provided the first measurements of the daily and seasonal behavior of water at the surface of Mars, including measurements of the abundance of adsorbed water in martian soil and near-surface ground ice at a high-latitude landing site. The absence of a plan to recover the MVACS investigation is a notable deficiency in NASA's present mission queue.

Despite its stated goal, "Follow the water," NASA's future Mars mission plans beyond those designed to recover past failures do not contain any investigations that involve actual water measurements. Most of the resources of the Mars program are being devoted to efforts to better characterize the mineralogy of rocks, look for sedimentary deposits, and search for landing sites and technologies that will be used for sample return.

The three-dimensional distribution of water in the martian crust is a critical measurement. As noted in this chapter, the observed reservoirs of water do not account for the amount of water expected based on cosmochemical arguments, nor the amount required to create geomorphic features such as outflow channels and valley networks. What we do see is thought to be the tip of the iceberg, but that has not been demonstrated. It is entirely possible that much of Mars's original inventory of water has escaped to space.[37]

The GRS on Mars Odyssey will map hydrogen in the near surface; however, GRS's low spatial resolution (300 km) and the fact that it only detects hydrogen in the top meter of soil make it only a first step in this effort. Deeper penetration and higher spatial resolution are required to characterize the water reservoirs. Future plans call for radar sounding to detect segregated ice and water deposits in the crust (e.g., the MARSIS radar experiment on Mars Express and the sounding radar planned for MRO). While of great promise, it is not yet clear that these observations will unambiguously resolve the nature of subsurface ice and water. A combination of orbital and landed science packages (some that may include drilling) will probably be required.

From a scientific standpoint, a more balanced water strategy in which missions designed to understand Mars's past water history are pursued in parallel with missions designed to understand the present behavior of water on Mars would be a more prudent approach. For example, loss of the MVACS experiment on MPL and the lack of plans to recover these measurements mean that there will be no direct measurements of water in the soil and how it exchanges with the atmosphere. There are no plans to directly measure water that is contained within the polar caps or perhaps sequestered in ground ice at lower latitudes. Direct measurement of isotopic ratios of atmospheric gases over a martian year would also provide constraints on the sources, sinks, and reservoirs of volatiles (see Chapters 8 and 12 in this report). Such measurements would provide essential data to significantly enhance our understanding of water and thus support the "Follow the water" strategy.

Recommendation. COMPLEX recommends that NASA pursue the global mapping of subsurface water and water ice in near-surface and crustal reservoirs.

REFERENCES

1. H.H. Kieffer and A.P. Zent, "Quasi-periodic Climate Change on Mars," pp. 1180–1220 in *Mars*, H.H. Kieffer, B.M. Jakosky, C.W. Snyder, and M.S. Matthews (eds.), University of Arizona Press, Tucson, 1992.
2. S.M. Clifford, D. Crisp, D.A. Fisher, K.E. Herkenhoff, S.E. Smrekar, P.C. Thomas, D.D. Wynn-Williams, R.W. Zurek, J.R. Barnes, B.G. Bills, E.W. Blake, W.M. Calvin, J.M. Cameron, M.H. Carr, P.R. Christensen, B.C. Clark, G.D. Clow, J.A. Cutts, D. Dahl-Jensen, W.B. Durham, F.P. Fanale, J.D. Farmer, F. Forget, K. Gotto-Azuma, R. Grard, R.M. Haberle, W. Harrison, R. Harvey, A.D. Howard, A.P. Ingersoll, P.B. James, J.S. Kargel, H.H. Kieffer, J. Larsen, K. Lepper, M.C. Malin, D.J. McCleese, B. Murray, J.F. Nye, D.A. Paige, S.R. Platt, J.J. Plaut, N. Reeh, J.W. Rice, Jr., D.E. Smith, C.R. Stoker, K.L. Tanaka, E. Mosley-Thompson, T. Thorsteinsson, S.E. Wood, A. Zent, M.T. Zuber, and H.J. Zwally, "The State and Future of Mars Polar Science and Exploration," *Icarus* 14: 210–242, 2000.
3. D.E. Smith, M.T. Zuber, S.C. Solomon, R.J. Phillips, J.W. Head, J.B. Garvin, W.B. Banerdt, D.O. Muhleman, G.H. Pettengill, G.A. Neumann, F.G. Lemoine, J.B. Abshire, O. Aharonson, C.D. Brown, S.A. Hauck, A.B. Ivanov, P.J. McGovern, H.J. Zwally, and T.C. Duxbury, "The Global Topography of Mars and Implications for Surface Evolution," *Science* 284: 1495–1503, 1999.

4. F.P. Fanale, J.R. Salvail, W.B. Banerdt, and R.S. Saunders, "Mars: The Regolith-Atmosphere-Cap System and Climate Change," *Icarus* 50: 381–407, 1982.
5. M.H. Carr, *Water on Mars*, Oxford University Press, New York, 1996.
6. A.P. Zent, F.P. Fanale, J.R. Salvail, and S.E. Postawko, "Distribution and State of H_2O in the High-Latitude Shallow Subsurface of Mars," *Icarus* 67: 19–36, 1986.
7. R.E. Arvidson, J.L. Gooding, and H.J. Moore, "The Martian Surface as Imaged, Sampled, and Analyzed by the Viking Landers," *Reviews of Geophysics* 27: 39–60, 1989.
8. L.L. Watson, I.D. Hutcheon, S. Epstein, and E.M. Stolper, "Water on Mars: Clues from Deuterium/Hydrogen and Water Contents of Hydrous Phases in SNC Meteorites," *Science* 265: 86–90, 1994.
9. P. Thomas, S. Squyres, K. Herkenhoff, A. Howard, and B. Murray, "Polar Deposits on Mars," pp. 767–798 in *Mars*, H.H. Kieffer, B.M. Jakosky, C.W. Snyder, and M.S. Matthews (eds.), University of Arizona Press, Tucson, 1980.
10. R.B. Leighton and B.C. Murray, "Behavior of Carbon Dioxide and Other Volatiles on Mars," *Science* 153: 136–144, 1966.
11. M.H. Carr, *Water on Mars*, Oxford University Press, New York, 1996.
12. D.A. Paige, "The Thermal Stability of Near-Surface Ground Ice on Mars," *Nature* 356: 43–45, 1992.
13. V.R. Baker, "Water and the Martian Landscape," *Nature* 412: 228–236, 2001.
14. S.M. Clifford, "A Model for the Hydrologic and Climatic Behavior of Water on Mars," *Journal of Geophysical Research* 98: 10973–11016, 1993.
15. V.C. Gulick, "Origin of Valley Networks on Mars: A Hydrological Perspective," *Geomorphology* 37: 241–268, 2001.
16. A.P. Ingersoll, "Mars: Occurrence of Liquid Water," *Science* 168: 972–973, 1970.
17. C.B. Farmer and P.E. Doms, "Global and Seasonal Variation of Water Vapor on Mars and Implications for Permafrost," *Journal of Geophysical Research* 84: 2881–2888, 1979.
18. B.M. Jakosky, "The Seasonal Cycle of Water on Mars," *Space Science Reviews* 41: 131–200, 1985.
19. M.D. Smith, J.C. Pearl, B.J. Conrath, and P.R. Christensen, "Recent TES Results: Mars Water Vapor Abundance and the Vertical Distribution of Aerosols," *Bulletin of the American Astronomical Society* 32: 3, 2000.
20. R.M. Haberle and B.M. Jakosky, "Sublimation and Transport of Water from the North Residual Polar Cap on Mars," *Journal of Geophysical Research* 95: 1423–1437, 1990.
21. M.I. Richardson, "The Water Cycle: Dynamics of Reservoir Exchange, Transport, and Integrated Behaviour," Paper presented at the Fifth International Conference on Mars, Pasadena, Calif., July 18–23, 1999.
22. S.M. Clifford, "A Model for the Hydrologic and Climatic Behavior of Water on Mars," *Journal of Geophysical Research* 98: 10973–11016, 1993.
23. M.H. Carr, *Water on Mars*, Oxford University Press, New York, 1996.
24. D.E. Smith, M.T. Zuber, S.C. Solomon, R.J. Phillips, J.W. Head, J.B. Garvin, W.B. Banerdt, D.O. Muhleman, G.H. Pettengill, G.A. Neumann, F.G. Lemoine, J.B. Abshire, O. Aharonson, C.D. Brown, S.A. Hauck, A.B. Ivanov, P.J. McGovern, H.J. Zwally, and T.C. Duxbury, "The Global Topography of Mars and Implications for Surface Evolution," *Science* 284: 1495–1503, 1999.
25. M.C. Malin and K.S. Edgett, "Sedimentary Rocks of Early Mars," *Science* 290: 1927–1937, 2000.
26. M.C. Malin and K.S. Edgett, "Evidence for Recent Groundwater Seepage and Surface Runoff on Mars," *Science* 288: 2330–2335, 2000.
27. C.B. Leovy, "Reconsidering Martian winds," Scientific American Explore, available online at <http://www.sciam.com article.cfm?articleID=0001696B-5AFC-1C75-9B81809EC588EF21&pageNumber=1&catID=4>, accessed April 21, 2003.
28. T.J. Parker, D.S. Gorsline, R.S. Saunders, D.C. Pieri, and D.M. Schneeberger, "Coastal Geomorphology of the Martian Northern Plains," *Journal of Geophysical Research* 98: 11061–11078, 1993.
29. J.W. Head III, H. Hiesinger, M.A. Ivanov, M.A. Kreslavsky, S. Pratt, and B.J. Thomson, "Possible Oceans in Mars: Evidence from Mars Orbiter Laser Altimeter Data," *Science* 286: 2134–2137, 1999.
30. J.F. Mustard, C.D. Cooper, and M.K. Rifkin, "Evidence for Recent Climate Change on Mars from the Identification of Youthful Near-Surface Ground Ice," *Nature* 412: 411–414, 2001.
31. V.R. Baker, "Water and the Martian Landscape," *Nature* 412: 228–236, 2001.
32. M.C. Malin and K.S. Edgett, "Sedimentary Rocks of Early Mars," *Science* 290: 1927–1937, 2000.
33. H.Y. McSween, Jr., "What We Have Learned About Mars from SNC Meteorites," *Meteoritics* 29: 757–779, 1994.
34. R.M. Haberle, H.C. Houben, and R.E. Young, "Multiannual Simulations with the Mars Climate Model," p. 14 in *Workshop on Atmospheric Transport on Mars*, Lunar and Planetary Institute, Houston, Texas, 1993.
35. F. Forget, F. Hourdin, R. Fournier, C. Hourdin, O. Talagrand, M. Collins, S. R. Lewis, P. L. Read, and J.-P. Huot, "Improved General Circulation Models of the Martian Atmosphere from the Surface to Above 80 km," *Journal of Geophysical Research* 104: 24155–24176, 1999.
36. C. Leovy, "Weather and Climate on Mars," *Nature* 412: 245–249, 2001.
37. B.M. Jakosky and R.J. Phillips, "Mars Volatile and Climate Evolution: Water the Real Constraints?," 32nd Lunar and Planetary Science Conference, Abstract #1147 (CD-ROM), 2001.

7

Life, Fossils, and Reduced Carbon

PRESENT STATE OF KNOWLEDGE

The search for evidence of extant or fossil martian life will focus on life forms that are carbon-based, that require water, and that are microbial. The implied assumptions are well justified. All known forms of life are carbon-based and require water. Throughout Earth's history, microbe-level life has been predominant—"simple" life forms can be expected to exist even if more advanced organisms have arisen subsequently. Moreover, it seems likely that life on Mars, if there is any, would take forms similar at some fundamental functional level to microorganisms on Earth, since the same laws of physics as they relate to biological energy transformations and to potentially habitable environments presumably apply on both planets. The subject of life detection in extraterrestrial samples is discussed in a recent study by the Space Studies Board.[1]

Microbes on Earth

Microbes are capable of surviving and growing under a broad range of environmental conditions (light intensity; total salinity; pH; temperature; oxygen abundance; carbon dioxide concentrations; water availability; and fluxes of ultraviolet, x-ray, gamma-ray, and highly ionizing radiation).[2] Environments inhabited by microbial assemblages are equally diverse—searing sabhkas;[3] antarctic lakes;[4] rock crusts of frozen deserts;[5] sea ice at –35° C (see Figure 7.1); solid salt crystals;[6] seafloor hydrothermal fluids;[7] and groundwaters and pore spaces of deep subsurface rocks.[8,9,10] In recognition of this environmental flexibility, it has been proposed that particular assemblages of microorganisms having specific physiological capabilities could survive on Mars.[11,12,13,14,15] Some of these proposals assume that the primary producers in such assemblages would be light-dependent (photosynthesizing) microbes for which survival at the martian surface may be precluded by harsh conditions. If there is no feasible photosynthetic zone on Mars, any extant life would have to obtain energy from inorganic sources. Such sources are used by terrestrial microbes in deep marine environments[16] and in the pore spaces of deep subsurface rocks,[17] but the detailed physiology of these microbes and the biological composition, spatial distribution, and ecology of the systems they live in have been incompletely investigated.

The present martian surface is oxidizing, desiccated, and bathed in intense ultraviolet radiation. It is possible, therefore, that the martian regolith (the pulverized rock debris that covers most of the surface) will prove to be uninhabitable by any living system and inimical to all but short-term survival of organic carbon.[18] Conditions less

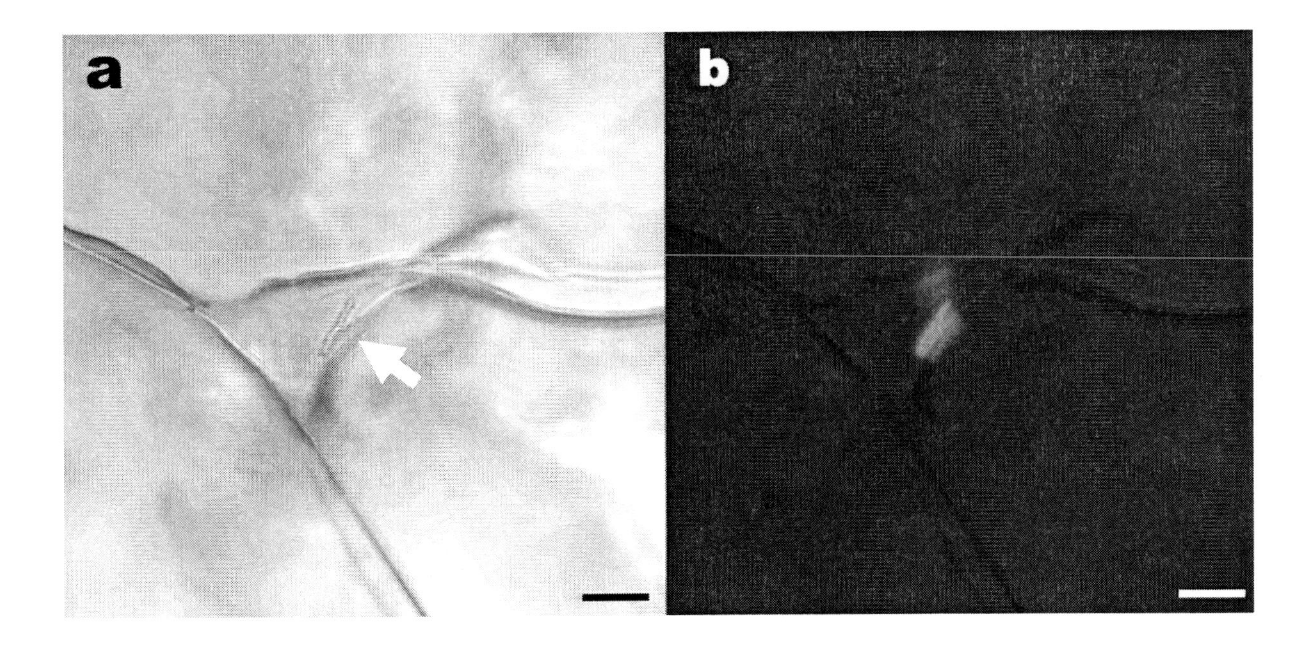

FIGURE 7.1 Inclusion of brine (triangular) at a triple junction between ice crystals in sea ice. Scale bars, 10 μm. (a) Transmitted light; arrow points to two rod-shaped bacteria along the wall of the inclusion. Higher magnification reveals apparent cell division. (b) Epifluorescent image of the same field; the bacteria are DAPI stained. Note the reflectance of the fluorescent emission from the organisms off the ice walls. SOURCE: K. Junge, C. Krembs, J. Deming, A. Stierle, and H. Eicken, "A Microscopic Approach to Investigate Bacteria Under In-Situ Conditions in Sea-Ice Samples," in *Selected Papers from the International Symposium on Sea Ice and Its Interactions with the Ocean, Atmosphere, and Biosphere, Fairbanks, Alaska, 18–23 June 2000*, M.D. Jeffries and H. Eicken (eds.), *Annals of Glaciology* 33: 304–310, 2001.

hostile to life may exist in sheltered habitats, and highly resistant spores or cysts dispersed by putative organisms occupying these or even more clement environments (such as the deep subsurface) might survive in the regolith. Moreover, as discussed below, the martian surface may have harbored life at an earlier time, when conditions were more favorable. If so, microscopic fossils of such organisms may be preserved in appropriately ancient sedimentary settings, and the evolutionary descendants of these microbes may inhabit the martian subsurface today. Indeed, if reports are valid of microbes surviving for 25 million to 40 million years encased in amber[19] and up to 100 million years encased in halite crystals,[20] it is conceivable that viable biological remnants might be harbored by sedimentary mineral precipitates on the martian surface. Mineral deposits of this sort also seem a promising site for the detection of inorganic microbial pseudomorphs (mineral structures precipitated as a result of microbial physiologic activity), just as they are on Earth in a number of settings.[21,22,23,24] Examples of especially ancient (Precambrian) terrestrial microfossils are shown in Figure 7.2.

The Search for Extant Life on Mars

Possible Abodes

The surface of Mars today is cold, dry, chemically oxidizing, and exposed to an intense flux of solar ultraviolet radiation. These four factors are likely to limit or even to prohibit life at or near the surface of the martian regolith.

Temperature is of interest not only because of its controlling influence on microbial metabolic rates, but also because of its influence on the stability of liquid water. Although the peak daytime surface temperature near the

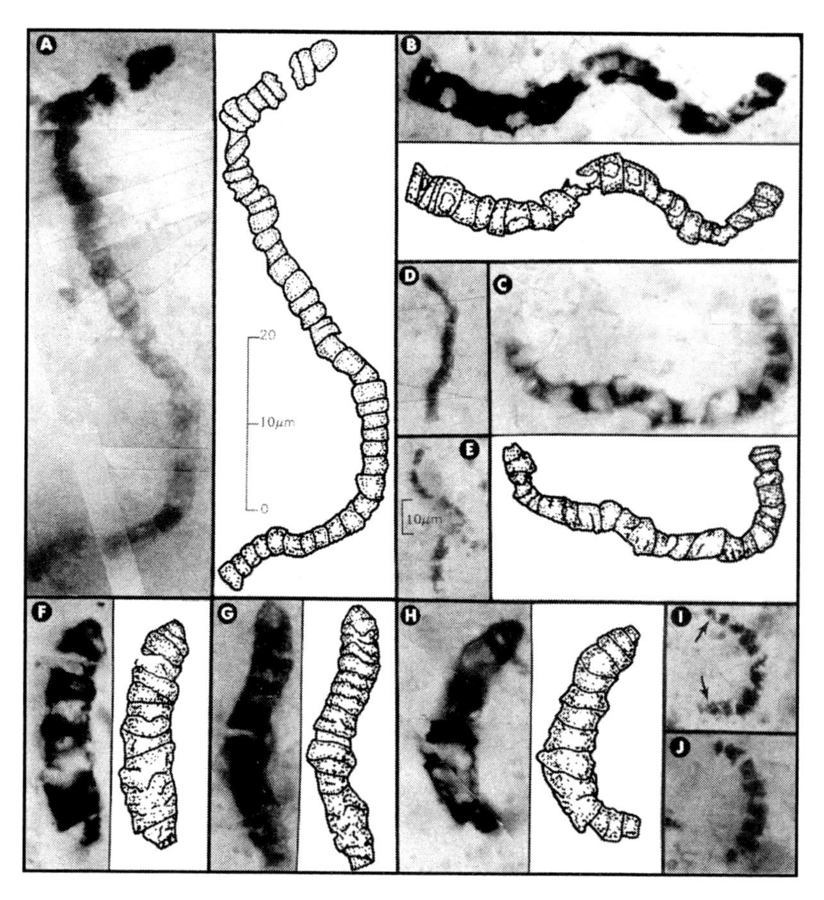

FIGURE 7.2 Carbonaceous prokaryotic fossil microorganisms, with interpretive drawings, in 3.46 billion-year-old siliceous rocks (chert) from Western Australia. All these fossils are completely encased in chert, which has been made into petrographic thin sections, rendering them visible under the microscope. The 10-μm scale bar in E is also applicable to D, I, and J; all other images are to the scale shown in A. SOURCE: Reprinted with permission from J.W. Schopf, "Microfossils of the Early Archean Apex Chert," *Science* 260: 640–646, 1993. Copyright 1993 by the American Association for the Advancement of Science. For additional information and an alternative interpretation see, for example, J.W. Schopf, A.B. Kudryavtsev, D.G. Agresti, T.J. Wdowiak, and A.D. Czaja, "Laser-Raman Imagery of Earth's Earliest Fossils," *Nature* 416: 73–76, 2002; and M.D. Brasier, O.R. Green, A.P. Jephcoat, A.K. Kleppe, M.J. Van Kranendonk, J.F. Lindsay, A. Steele, and N.V. Grassineau, "Questioning the Evidence for Earth's Oldest Fossils," *Nature* 416: 76–81, 2002.

martian equator can rise above the freezing point of water during much of the year, the average surface temperature is about –55° C, well below the freezing point of water.

All life on Earth is based on aqueous chemistry—liquid water is essential for life as we know it. Water is abundant on Mars (see Chapter 6 in this report), but not in liquid form.[25] Water vapor and ice crystals are present in the atmosphere, and water ice is almost certainly present within the martian regolith at high latitudes and at the surface in polar regions. During the half-year-long, north-polar summer, the water ice present in the residual polar cap warms sufficiently to allow water to sublime into the atmosphere and be distributed globally. Because dissolution of salts can lower the freezing point, it is possible that liquid water exists transiently on or near the martian surface. Such occurrences are probably very rare, since such saline liquids would rapidly evaporate.[26] Alternatively, ice crystals trapped in closed pores in rocks or regolith grains could melt under certain circumstances, and the resulting liquid water could be prevented from evaporating by virtue of being enclosed. At increasing depth, as subsurface temperatures rise as a result of the planetary geothermal gradient, liquid water may be present in pore spaces.[27,28]

The evidence that the surface environment is highly oxidizing derives from analyses by instruments carried aboard the 1976 Viking landers. Although the exact nature of the oxidants has not been determined,[29] the most likely dominant oxidant appears to be hydrogen peroxide, believed to form photochemically from atmospheric water vapor and to diffuse readily into the regolith. Such oxidants would react with organic molecules (including those that make up microbes) and may be responsible for the reported absence of organic molecules or their fragments in the martian samples analyzed by the Viking landers, despite the fact that organic material has been continually added to the regolith over planetary history by the impacts of carbonaceous meteorites.[30]

The martian atmosphere is thin, having an average pressure of ~6 mbar, and consists primarily of carbon dioxide. Because of the low concentration of atmospheric ozone, solar ultraviolet light reaches the surface of Mars with much less attenuation than on Earth, which is shielded by an ozone layer in its atmosphere. On Mars, winter-hemisphere atmospheric ozone can absorb some of the impinging ultraviolet radiation, but only during a brief period of the year and only over a small fraction of the planet. Thus, the entire martian surface is subject to an intense flux of ultraviolet radiation. Unless putative martian microbes were in some way shielded by ultraviolet-absorbing carapaces, composed of remarkably ultraviolet-stable organic compounds, and/or possessed intracellular molecular repair mechanisms decidedly more effective than those present in known terrestrial forms of life, any martian microbes would be expected to survive only in habitats protected from such radiation.

Status of the Search

The accepted interpretation of results from the Viking landers is that the surface materials tested were devoid of organic molecules and of any other evidence of life.[31] However, even without consideration of alternative interpretations,[32] the Viking results cannot be taken as indicating that life does not currently exist on Mars. Organisms at the Viking sites might have been missed because the experimental conditions (nutrients provided, processes followed) may not have been chosen correctly. Even more importantly, martian life might reside in aqueous oases, such as any recently active volcanic vents or fumaroles distant from the Viking landing sites, or at depths far beneath the surficial regolith sampled by the Viking experiments. Dormant life could exist for a time in dry settings, such as the interiors of rock fragments excavated from depth by cratering events.

The Search for Past Life on Mars

Possible Sources

The surface environment of Mars may not always have been as hostile to life as it is today. Early in the planet's history, the average temperature may have been warmer and the atmosphere more dense, and liquid water may have existed at the surface.

Evidence for the presence of surface water early in martian history is preserved in the geomorphology of the planet's surface (see Chapter 5 in this report), particularly in areas interpreted on the basis of the distribution and density of meteoritic impact craters to be older than about 3.5 billion years. Two aspects of this particularly ancient terrain suggest that the early climate differed substantially from that of the present.[33] First, large, early-formed craters have been nearly obliterated or substantially degraded in a style regarded to be indicative of aqueous erosion. Later-formed craters have survived essentially intact, suggesting that rates of erosion early in martian history may have been as much as 1,000 times greater than those of more recent epochs. Second, the pre-3.5 billion-year-old surface exhibits networks of dendritic valleys that have geomorphologic patterns closely similar to those of terrestrial stream channels carved by water. Regardless of the exact mechanism of formation of these networks (whether by runoff of precipitation, seepage or "sapping" of subsurface water, or erosion by water-rich debris flows or catastrophic floods), the valleys provide strong evidence of the presence of liquid water at or very near the martian surface.[34] Overall, the geomorphologic evidence indicates that the martian climate was wetter, warmer, and appreciably more hospitable to life prior to about 3.5 billion years ago than it is at present.

Hot-spring environments, another setting habitable by microbial forms of life, may also have been present, perhaps abundantly, on early Mars.[35] The formation of hot springs and hydrothermal systems requires the co-occurrence of crustal water and substantial heat. Meteoritic impacts, particularly common early in martian history, would have provided a strong source of local heating. Moreover, isotopic evidence from martian (SNC) meteorites indicates that Mars, like Earth, underwent crustal differentiation and, probably, global melting, during and shortly after planetary accretion. Thus, volcanism, no doubt especially extensive in early martian history (and possibly continuing to the near-present[36]), would have served as a second major heat source. Given the extensive evidence both of crustal water and of strong heat sources, biologically habitable hot springs and hydrothermal systems are likely to have been common on ancient Mars (and perhaps to have persisted throughout martian history).

A third widespread zone that seems likely to have been habitable throughout martian history is the crustal subsurface, where water may exist in a liquid state. The geothermal gradient of Mars is probably such that liquid water is present at depths as shallow as 2 km near the equator.[37] The discovery of terrestrial microbes living deep within the Columbia River basalts in the Pacific Northwest of the United States[38,39] and elsewhere on Earth, at depths as great as 3 km,[40] is consistent with the possible presence of microbes living in similar settings on Mars. On Earth, such microbes survive by metabolizing hydrogen produced by chemical reactions between pore water and the enclosing rocks. Although they are thought to be completely independent of the food chain and chemical gradients established by photosynthetic organisms at Earth's surface, they presumably migrated into such deep subsurface settings from elsewhere rather than having originated in place. If life was present at or near the surface of Mars early in its history, when the surface environment must have been wetter, warmer, and biologically more benign than at present, a similar migration to the subsurface might have occurred as the martian surface became increasingly less hospitable.

To summarize, fossil evidence of past martian life, if there is any, may be preserved in surface water-laid deposits such as lake or streambed sediments, in evaporitic mineral pans,[41] and in hydrothermally deposited mineral crusts. Subsurface settings are not as readily accessible, but materials from them might have been dislodged and brought to the surface by meteoritic impacts.

Status of the Search

Within the past few years, results of detailed studies have been interpreted as indicating that martian (SNC) meteorite ALH84001 contains possible evidence of biological activity thought to date from about 3.6 billion years ago.[42] The possibly biological indicators detected include mineral-zoned carbonate disks and grains of magnetite similar to those interpreted in some terrestrial deposits as being of biological origin, polycyclic aromatic hydrocarbons interpreted to be possible remnants of decayed and geochemically altered biological organic matter, and globular or filamentous microscopic structures thought possibly to be microbial fossils. These claims concerning ALH84001[43] have engendered much discussion, both pro and con, regarding each of the several intriguing indicators proposed.[44] To date, none has been shown to be decisively of biological origin. But the widespread interest stirred by the report illustrates the importance attached to the question of the past or present existence of life on Mars, by both the scientific community and the public at large.

It is important to maintain as the focus of the search for life the question, Did life ever arise on Mars? This is a broad and complex question, and the evidence may be so deeply buried in the past that the question can be answered only by acquiring an extensive and deep knowledge of Mars. For example, on Earth, enzyme-driven metabolic processes can create characteristic biogenic isotopic signatures (affecting, in particular, the ranges of compositions of the stable isotopes of carbon, sulfur, nitrogen, hydrogen, and possibly iron). However, in order to use isotopic measurements to test for the past presence of life, one needs to know the scope of abiotic fractionating processes. The search for life should be based on this premise, and one should be as prepared for a negative answer as for a positive one. The importance of a positive answer is clear, but a negative answer would prompt inquiries into what the implications are for the planetary differences between Earth and Mars.

NEAR-TERM OPPORTUNITIES

The search for evidence of present or past life on Mars is a classic hunt for a "needle in a haystack." Accordingly, the appropriate strategy is to first identify the "haystack"—promising sedimentary deposits and mineral accumulations, concentrations of reduced carbon—and only later to expect to find the "needle" that may be hidden therein: living microbes, fossil microbes, or diagnostic organic compounds and isotopic compositions. In any case, the context for the needle and the haystack must be well understood through a balanced exploration and science program. The first sample-return mission will be the start of this process. Further missions are likely to extend for a half century or more. Success must be measured not in terms of finding the needle but in terms of steadily increasing our understanding of Mars and whether it could ever have had its own biota.

Robotic Exploration

Analyses from orbit hold promise for identifying biologically or paleontologically favorable sites for sample collection—lake or streambed sediments, evaporitic mineral pans, and hydrothermal deposits, hot springs, and similar locales.

The Mars Exploration Rovers, scheduled for launch in 2003, and the Mars Science Laboratory (MSL),[a] scheduled for launch in 2007,[b] will contribute strongly to an overall understanding of the Mars system and thus provide vital information about the context for life and for prebiotic chemical processes. These are very significant points that will help to refine thinking about biological issues, but no decisive revelations about martian life can be expected from these missions. Especially if equipped with means for appropriate chemical analyses and imaging at the proposed 30-μm resolution, the rovers hold promise for useful in situ analyses of evaporitic mineral crusts and finely layered hydrothermal and other sedimentary deposits, as well as a search for concentrations of reduced carbon. The U.K.'s Beagle 2, scheduled for launch in 2003 aboard the European Space Agency's Mars Express, is most promising in this regard. Uniquely among the landed missions and rovers, it has the potential to provide results that bear directly and dramatically on the life sciences at the same time that it contributes strongly to important planetological questions.

Rovers will come to have the capability of drilling, but it is questionable whether they can ever reach beneath the surface zone of Mars in which conditions are particularly hostile to life and apparently to the survival of organic materials. Moreover, although any cores collected can be studied robotically, it would be far more profitable to return them to Earth for study.

Returned Samples

The capabilities of robotic instruments that can be launched from Earth and flown to Mars are dwarfed by those of Earth-bound laboratories. Diverse physical and chemical characteristics of any returned sample of martian soil can be analyzed grain by grain. The results will provide information not just about a particular sample of soil, but about all of the rocks that had been weathered to provide that soil, about the fluids with which those grains had come in contact, and about the processes by which the grains had reached the site of collection. The resulting picture will provide information about the region from which the samples are collected and about the history of the whole martian environment. Robotic missions have much to offer but, in terms of the search for life or for evidence of past life, access to returned samples is incomparably superior to remote analyses.

---

[a]MSL is also referred to as the Mars Smart Lander, the Mobile Surface Laboratory, and by a variety of other names.

[b]Following the completion of this study, NASA announced that it was delaying the launch of MSL until 2009 to allow time to develop an advanced, radioisotope power system for this mission.

RECOMMENDED SCIENTIFIC PRIORITIES

Earlier studies of Mars exploration strategy (see Appendix B: [1.7, 1.19, 3.3, 5, 6.2, 7, 9.1, 10.3, 11.3.1]), reinforced by the establishment and importance of the Astrobiology program in NASA, have given very high priority to the search for life, extant or fossil, on that planet. Among the earlier recommendations are these:

- From the 1990 COMPLEX report *1990 Update to Strategy for Exploration of the Inner Planets*:[45] "Consistent with the SSB report *The Search for Life's Origins: Progress and Future Directions in Planetary Biology and Chemical Evolution* (National Academy Press, Washington, D.C., 1990), the committee endorses the continued search for evidence of past life and biochemical evolution on Mars, as well as the continuing study of the history of water on Mars" (Appendix B: [3.3]).
- From the 1998 COMPLEX letter report "Assessment of NASA's Mars Exploration Architecture":[46] "An appropriate focus for NASA's Mars program is the comprehensive goal of understanding Mars as a possible abode of past or present life" (Appendix B: [9.1]). The wording of this recommendation is highly significant. *Understanding Mars* as a possible abode of life will be a stepwise process. Mandatory and fundamental steps are the establishment of a geochronologic framework for martian surface deposits, reconstruction of the history of volatiles at the martian surface in as much detail as possible, and development of a good overview of the geologic processes affecting the martian surface. As that work proceeds, it will be possible to define new objectives more tightly focused on biological processes and products. Returning to the needle-in-a-haystack analogy, first one must find the right fields, then select the most promising haystacks, and then search for the needle.
- From the 2000 MEPAG report "Mars Exploration Program: Scientific Goals, Objectives, Investigations, and Priorities":[47] "Explore high priority candidate sites (i.e., those that provide access to near-surface liquid water) for evidence of extant (active or dormant) life" (Appendix B: [11.3.1]).

ASSESSMENT OF PRIORITIES IN THE MARS EXPLORATION PROGRAM

NASA has made the search for life, or evidence of it, central to its program of Mars exploration. The definitive detection of life on Mars and its study will almost certainly require the return of samples to Earth (see Chapter 11 in this report). Unfortunately, in response to mission failures in 1999 and to the ensuing restructuring of its plans for Mars exploration, NASA has deemed it necessary to defer the launch of the first sample-return mission by at least 8 years, from 2003 when the first launch had originally been planned to 2011 under the Mars Exploration Program schedule.

The priorities recommended by advisory panels, including those quoted above, can be summarized by what has come to be an often-cited NASA slogan: "Seek, In Situ, Sample." The first two elements of this slogan are being pursued, in the form of the robotic orbiter and lander missions described throughout this report, and COMPLEX judges that the planned effort is sufficient to set the stage for rational choice of a landing site in 2011 (see Chapter 12 in this report). Regarding the third element of the slogan, little has been done except to commit to it in principle. NASA has taken few concrete actions in support of sample-return missions. COMPLEX recommends that the following important steps be taken (see also Chapter 12).

Recommendations

- The Mars Quarantine Facility in which martian samples will be processed, stored, and released for scientific study, and in which a very limited range of studies will be carried out, must be designed, built, and certified.[48]
- Research must be initiated on several outstanding questions that will affect the design of the Mars Quarantine Facility (e.g., combining biological isolation with clean-room conditions; establishing the efficacy and detrimental effects of sterilization techniques).[49]
- The study of life in extreme environments on Earth, which can aid in the design of life-detection tests, should be supported, as is already being done. In general, research areas that improve sensitivities of life-

detection techniques must be supported, and a life-detection protocol to be implemented and tested in the Mars Quarantine Facility must be developed.
- Techniques must be developed for the collection, packaging, and return of samples.
- The research programs of Mars orbiter and lander missions must be designed to support the collection of those samples with the greatest potential for life detection (this process is underway).

NASA should focus its Mars program, and sample-return missions in particular, on the comprehensive goal of understanding Mars as a possible abode of life. That is, the interpretation of biologically relevant in situ and laboratory observations can be maximized only if data are gathered in the context of a broad framework of research aimed at understanding the origin and evolution of the martian environment.[50]

REFERENCES

1. Space Studies Board, National Research Council, *Signs of Life: A Report Based on the April 2000 Workshop on Life Detection Techniques*, National Academy Press, Washington, D.C., 2002.
2. See, for examples, references 33 through 61 in J.W. Schopf, "Disparate Rates, Differing Fates: Tempo and Mode of Evolution Have Changed from the Precambrian to the Phanerozoic," *Proceedings of the National Academy of Sciences* 91: 6735–6742, 1994.
3. B.K. Pierson, "Modern Mat-Building Microbial Communities: Introduction," pp. 247–251 in *The Proterozoic Biosphere—A Multidisciplinary Study*, J.W. Schopf and C. Klein (eds.), Cambridge University Press, New York, 1992.
4. B.C. Parker, G.M. Simmons, Jr., G. Love, R.W. Wharton, and K.G. Seaburg, "Modern Stromatolites in Antarctic Dry Valley Lakes," *Bioscience* 31: 656–661, 1981.
5. E.I. Freidman and R. Ocampo-Freidman, "Endolithic Microorganisms in Extreme Dry Environments: Analysis of a Lithobiotic Microbial Habitat," pp. 177–185 in *Current Perspectives in Microbiology*, M.J. Klug and C.A. Reddy (eds.), American Society of Microbiology, Washington, D.C., 1984.
6. C.F. Norton, T.J. McGenity, and W.D. Grant, "Archeal Halophiles (Halobacteria) from Two British Salt Mines," *Journal of General Microbiology* 139:1077–1081, 1993.
7. H.W. Jannasch, "Microbial Interactions with Hydrothermal Fluids," pp. 273–296 in *Seafloor Hydrothermal Systems: Physical, Chemical, Biological, and Geological Interactions*, Geophysical Monograph 91, American Geophysical Union, Washington, D.C., 1995.
8. T.O. Stevens and J.P. McKinley, "Lithoautotrophic Microbial Ecosystems in Deep Basalt Aquifers," *Science* 270: 450–454, 1995.
9. T.O. Stevens, "Lithoautotrophy in the Subsurface," in *Proceedings of the Third International Symposium of Subsurface Microbiology*, Sept. 15–21, 1996, Davos, Switzerland, Swiss Society of Microbiology, Zurich, 1996.
10. S. Kostelnikova and K. Pederson, "Ecology of Methanogenic Archea in Granitic Groundwater from Hard Rock Laboratory, Sweden," in *Proceedings of the Third International Symposium of Subsurface Microbiology*, Sept. 15–21, 1996, Davos, Switzerland, Swiss Society of Microbiology, Zurich, 1996.
11. E.I. Freidman and R. Ocampo-Freidman, "Endolithic Microorganisms in Extreme Dry Environments: Analysis of a Lithobiotic Microbial Habitat," pp. 177–185 in *Current Perspectives in Microbiology*, M.J. Klug and C.A. Reddy (eds.), American Society of Microbiology, Washington, D.C., 1984.
12. C.P. McKay, E.I. Freidman, R.A. Wharton, Jr., and W.L. Davies, "History of Water on Mars: A Biological Perspective," *Advances in Space Research* 12: 231–238, 1992.
13. C.P. McKay, R.L. Mancinelli, C.R. Stoker, and R.A. Wharton, Jr., "The Possibility of Life on Mars During a Water-Rich Past," pp. 1234–1245 in *Mars*, H.H. Kieffer, B.M. Jakosky, C.W. Snyder, and M.S. Matthews (eds.), University of Arizona Press, Tucson, 1992.
14. P.J. Boston, M.V. Ivanov, and C.P. McKay, "On the Possibility of Chemosynthetic Ecosystems in Subsurface Habitats on Mars," *Icarus* 95: 300–330, 1992.
15. T.O. Stevens and J.P. McKinley, "Lithoautotrophic Microbial Ecosystems in Deep Basalt Aquifers," *Science* 270: 450–454, 1995.
16. See, for example, H.W. Jannasch, "Microbial Interactions with Hydrothermal Fluids," pp. 273–296 in *Seafloor Hydrothermal Systems: Physical, Chemical, Biological, and Geological Interactions*, Geophysical Monograph 91, American Geophysical Union, Washington, D.C., 1995.
17. See, for example, T.O. Stevens and J.P. McKinley, "Lithoautotrophic Microbial Ecosystems in Deep Basalt Aquifers," *Science* 270: 450–454, 1995.
18. J. Oró and G. Holzer, "The Photolytic Degradation and Oxidation of Organic Compounds Under Simulated Martian Conditions," *Journal of Molecular Evolution* 14:153–160, 1979.
19. R.J. Cano and M.K. Boruncki, "Revival and Identification of Bacterial Spores in 25- to 40-Million-Year-Old Dominican Amber," *Science* 268: 1060–1064, 1995.

20. C.F. Norton, T.J. McGenity, and W.D. Grant, "Archeal Halophiles (Halobacteria) from Two British Salt Mines," *Journal of General Microbiology* 139:1077–1081, 1993.
21. T.J. Beveridge, J.D. Meloche, W.S. Fyfe, and R.G.E. Murray, "Diagenesis of Metals Chemically Complexed to Bacteria: Laboratory Formation of Metal Phosphates, Sulfides, and Organic Condensates in Artificial Sediments," *Applied Environmental Microbiology* 45: 1094–1108, 1983.
22. F.G. Ferris, R.G. Wiese, and W.S. Fyfe, "Precipitation of Carbonate Minerals by Microorganisms: Implications for Silicate Weathering and the Global Carbon Dioxide Budget," *Geomicrobiology Journal* 12: 1–13, 1994.
23. G. Southam and T.J. Beveridge, "The In-Vitro Formation of Placer Gold by Bacteria," *Geochimica et Cosmochimica Acta* 58: 4527–4530, 1994.
24. G. Southam, F.G. Ferris, and T.J. Beveridge, "Mineralized Bacterial Biofilms in Sulphide Tailings and in Acid Mine Drainage Systems," pp. 148-170 in *Microbial Biofilms*, H.M. Lappinscott and J.W. Costerston (eds.), Cambridge University Press, Cambridge, U.K.
25. See, for example, B.M. Jakosky and R.M. Haberle, "The Seasonal Behavior of Water on Mars," pp. 969–1016 in *Mars*, H.H. Kieffer, B.M. Jakosky, C.W. Snyder, and M.S. Matthews (eds.), University of Arizona Press, Tucson, 1992.
26. B.C. Clark and D.C. Van Hart, "The Salts of Mars," *Icarus* 45: 370–378, 1981.
27. M.H. Carr, *Water on Mars*, Oxford University Press, New York, 1996.
28. See also the committee's reservations about the distribution of groundwater, in Chapter 6 of this report.
29. D.M. Hunten, "Possible Oxidant Sources in the Atmosphere and Surface of Mars," *Journal of Molecular Evolution* 14: 71–78, 1979.
30. K. Biemann, J. Oró, P. Toulmin III, L.E. Orgel, A.O. Nier, D.M. Anderson, P.G. Simmonds, D. Flory, A.V. Diaz, D.R. Rushneck, J.E. Biller, and A.L. Lafleur, "The Search for Organic Substances and Inorganic Volatile Compounds in the Surface of Mars," *Journal of Geophysical Research* 82: 4641–4658, 1977.
31. H.P. Klein, "The Viking Mission and the Search for Life on Mars," *Reviews of Geophysics and Space Physics* 17: 1655–1662, 1979.
32. See, for example, G.V. Levin, and P.A. Straat. "Viking Labeled Release Biology Experiment: Interim Results," *Science* 194: 1322–1329, 1976.
33. S.W. Squyres and J.F. Kasting, "Early Mars: How Warm and How Wet?" *Science* 265: 774–779, 1994.
34. M.H. Carr, *Water on Mars*, Oxford University Press, New York, 1996.
35. G.R. Brakenridge, H.E. Newsom, and V.R. Baker, "Ancient Hot Springs on Mars: Origin and Paleoenvironmental Significance of Small Martian Valleys," *Geology* 13: 859–862, 1985.
36. R. Greeley and B.D. Schneid, "Magma Generation on Mars: Amounts, Rates, and Comparisons with Earth, Moon, and Venus," *Science* 254: 996–998, 1991.
37. M.H. Carr, *Water on Mars*, Oxford University Press, New York, 1996.
38. T.O. Stevens and J.P. McKinley, "Lithoautotrophic Microbial Ecosystems in Deep Basalt Aquifers," *Science* 270: 450–454, 1995.
39. T.O. Stevens, "Lithoautotrophy in the Subsurface," in *Proceedings of the Third International Symposium of Subsurface Microbiology*, Sept. 15–21, 1996, Davos, Switzerland, Swiss Society of Microbiology, Zurich, 1996.
40. See, for example, S. Kostelnikova and K. Pederson, "Ecology of Methanogenic Archea in Granitic Groundwater from Hard Rock Laboratory, Sweden," in *Proceedings of the Third International Symposium of Subsurface Microbiology*, Sept. 15–21, 1996, Davos, Switzerland, Swiss Society of Microbiology, Zurich, 1996.
41. J.L. Gooding, "Soil Mineralogy and Chemistry on Mars: Possible Clues from Salts and Clays in SNC Meteorites, *Icarus* 99: 28–41, 1992.
42. D.S. McKay, E.K. Gibson, Jr., K.L. Thomas-Keprta, H. Vali, C.S. Romanck, S.J. Clemett, X.D.F. Chillier, C.R. Macchling, and R.N. Zare, "Search for Past Life on Mars: Possible Relic Biogenic Activity in Martian Meteorite ALH84001," *Science* 273:924–930, 1996.
43. D.S. McKay, E.K. Gibson Jr., K.L. Thomas-Keprta, H. Vali, C.S. Romanck, S.J. Clemett, X.D.F. Chillier, C.R. Macchling, and R.N. Zare, "Search for Past Life on Mars: Possible Relic Biogenic Activity in Martian Meteorite ALH84001," *Science* 273:924–930, 1996.
44. A. Treiman, "On the Question of the Martian Meteorite: Recent Scientific Papers on ALH 84001 Explained, with Insightful and Totally Objective Commentaries," available online at <http://cass.jsc.nasa.gov/lpi/meteorites/alhnpap.html>, 2000.
45. Space Studies Board, National Research Council, *1990 Update to Strategy for Exploration of the Inner Planets*, National Academy Press, Washington, D.C., 1990.
46. Space Studies Board, National Research Council, "Assessment of NASA's Mars Exploration Architecture," letter report to Carl Pilcher, NASA, November 11, 1998.
47. NASA, Mars Exploration Payload Assessment Group (MEPAG), "Mars Exploration Program: Scientific Goals, Objectives, Investigations, and Priorities," December 2000, in *Science Planning for Exploring Mars*, JPL Publication 01-7, Jet Propulsion Laboratory, Pasadena, Calif., 2001.

48. Space Studies Board, National Research Council, *Quarantine and Certification of Martian Samples*, National Academy Press, Washington, D.C., 2002.
49. Space Studies Board, National Research Council, *Quarantine and Certification of Martian Samples*, National Academy Press, Washington, D.C., 2002.
50. Space Studies Board, National Research Council, "Assessment of NASA's Mars Exploration Architecture," letter report to Carl Pilcher, NASA, November 11, 1998.

8

Lower Atmosphere and Meteorology

PRESENT STATE OF KNOWLEDGE

This chapter describes the present-day atmosphere of Mars and briefly relates what it may tell about the atmosphere's past and future. Climate variability is covered in Chapter 9. The evolutionary course of Mars's atmosphere to its present state is uncertain, because its history is intimately tied to the crustal composition of the planet (see Chapter 3), and until a fairly detailed knowledge of surface materials is achieved, there cannot be confidence that atmospheric models attempting to duplicate the past or predict the future are correct. A large literature exists on isotopic analysis of the volatiles found in SNC meteorites as it relates to the history of water on Mars and the volatile inventory on the planet,[1] but efforts to infer the latter are severely hampered by ignorance of the surface and subsurface composition. For example, the discovery of large volumes of carbonate rock would profoundly influence our perception of the past and present state of the atmosphere.

Surface topography and geomorphology (see Chapter 5) indicate that active volcanism has expelled gases into the atmosphere. The identification of regions of high-feldspar basalt by the Thermal Emission Spectrometer (TES) on Mars Global Surveyor (MGS) points to basaltic volcanism and a basaltic surface composition.[2] Dust, however, covers much of the planet, and it could be masking the signature of other rock types. Patches of crystalline gray hematite have also been found,[3] but the extent and source of these are not certain, so the observation is not useful to atmospheric modeling.

Our knowledge of the composition of the Mars atmosphere is based on measurements of minor gases such as Ne, Kr, and Xe, and ratios of common isotopes in the ambient atmosphere ($^{36}Ar/^{38}Ar$, $^{12}C/^{13}C$, $^{16}O/^{17}O$, $^{16}O/^{18}O$, $^{14}N/^{15}N$, $^{2}H/^{1}H$) by the Viking descent mass spectrometer,[4] ground-based and airborne spectroscopy,[5] and laboratory analysis of atmospheric gases captured in the vitreous components of martian meteorites.[6] It is thought that a combination of impact erosion and long-term atmospheric loss from the top of the atmosphere by solar wind sputtering and other processes,[7,8] and possibly sequestration of CO_2 and other gases in the crust of the planet, are responsible for the present low atmospheric pressure at the surface of Mars (yearly average ~6 mbar) relative to pressures on Earth and Venus.

Mars's present-day lower atmosphere is dominated by the behavior of CO_2, water vapor, and dust, as driven by the Mars/Sun configuration, and their interactions with the surface. These factors, combined with issues of transport and cloud physics, comprise Mars meteorology. Seasonal changes in the atmospheric mass of CO_2 are up to 30 percent in the current epoch. Water vapor also interchanges with clouds and surface materials; its average annual column abundance is ~10 to 40 precipitable microns of H_2O at north midlatitudes.

CO_2 photolysis by sunlight at wavelengths shorter than 2275 Å would quickly convert the atmosphere to CO and O if there were not an efficient mechanism of recombination. Such a mechanism was predicted by McElroy and Donahue and by Parkinson and Hunten.[9,10] Each group postulated an odd-hydrogen (H, OH, HO_2) catalytic cycle that breaks the O_2 bond and provides an OH reservoir, and a pathway for the recombination of CO and O to form CO_2. Odd hydrogen is a natural consequence of photolysis of water vapor, which has been measured by orbital and ground-based instrumentation. Parkinson and Hunten showed that chemical breakup of H_2O_2 into two OH molecules would increase efficiency of the catalytic recombination cycle with no need for unrealistically large vertical mixing in the lower atmosphere.[11] The lack of in situ odd-hydrogen measurements constitutes a serious deficit in current knowledge of Mars's atmospheric photochemistry. There has never been an actual measurement of any odd-hydrogen compound, or H_2O_2, in Mars's lower atmosphere—still unknown are the major odd-hydrogen compounds, their lifetimes, and their precipitation character. The odd-hydrogen recombination theory is generally accepted because it has passed three major tests by predicting (1) the H abundance at the homopause that was observed by Mariner 9's Ultraviolet Imaging Photometer in Lyman-α airglow measurements and (2) the correct amount of O_3 and the way O_3 varies with moisture (season and latitude); and by explaining (3) the lack of living material at Viking landing sites (a result of highly oxidizing conditions, which do not favor the survival of organic compounds).

No orbital water vapor mapping has been made at wavelengths free of the influence and ambiguities introduced by airborne dust. While there is a good grasp of the global seasonal behavior of water vapor during the current epoch because of Viking's Mars Atmospheric Water Detector and Mars Global Surveyor TES measurements, knowledge of the daily variations of water vapor (i.e., measured at one place at different times of day) are scant and come mainly from ground-based observations.[12,13,14,15,16]

Actual measurements of O_2 in Mars's lower atmosphere have been made only from Earth, with coarse (regional) spatial discrimination. Earth-based observations have made and will continue to make valuable measurements, but they cannot achieve detailed spatial coverage. They do have the advantage of making simultaneous measurements at many longitudes and thus local times or geographic locations. The most detailed measurements of O_3 and O in the upper atmosphere are those made by Mariner 9 in the early 1970s,[17] when it was found that the O_3 abundance fluctuates inversely with that of water vapor and thus is greatest over the northern winter pole. Some measurements indicate fluctuations with topography and location. Thus, for a complete reconnaissance of these major constituents and photochemistry, measurements should be made from Mars's surface at a variety of topographic locations.

Dust raised from the surface into Mars's atmosphere strongly perturbs clean-air dynamics; prominent dust storms have been observed by ground-based telescopes, the Hubble Space Telescope, and orbiting spacecraft. Extensive imaging by Viking,[18] the Imager for Mars Pathfinder (IMP),[19] and the MGS Mars Orbiter Camera[20] has allowed great progress to be made in identifying the dust-loading mechanisms and quantifying the mass of airborne material. Important achievements allowed by the IMP measurements were more detailed characterization of airborne dust, and imaging of the blue aureole around the Sun. Other radiance measurements enabled quantification of the dust-particle size distribution, settling time, and strong phase-angle color effects. Results were in agreement with a reanalysis of the Viking data.[21] A better understanding of color differences resulting from the solar phase angle when using reflectance spectroscopic and multicolor imaging by remote sensing helps in applying corrections when making composition measurements.[22]

A more thorough understanding of the effect of dust loading on atmospheric temperature was achieved during the highly successful aerobraking of MGS and the science mapping measurements made by the spacecraft's TES. Temperature profiles for most latitudes and seasons were also obtained.[23] These, together with newly measured seasonal and spatial tidal amplitudes,[24] can, when fully analyzed, be synthesized into Mars global climate/circulation models, which will greatly improve predictions of Mars's atmospheric circulation and other details of Mars meteorology.

General Circulation and Seasonal Cycles

There now exists a basic understanding of the pole-to-pole circulation patterns that are responsible for seasonal transport of volatiles on Mars. Deep atmospheric profiles measured during the radio occultation of Mariner 9 were inverted,[25] and temperature and dynamical information was extracted. Some measurements made with instrumentation on Mars Pathfinder augment those made by the two Viking landers, but only at a few locations. The measurements did permit better modeling of global circulation and meteorology.[26] In addition, data assimilation techniques with modern global climate/circulation models using large data sets such as that from MGS's TES may permit conversion of limited coverage to full dynamical fields. However, because of possible coupling between global dust storms and volatile transport, it is difficult to estimate the long-term transport efficiency of "normal" circulation.

There are basically three regimes in Mars circulation: (1) polar condensed flow, (2) midlatitude baroclinic wave traveling weather systems, and (3) tropical Hadley cells. The Hadley circulation is associated with "trade winds" blowing from the northeast in the northern hemisphere and the southeast in the southern hemisphere. During summer and winter, martian trade winds cross the equator. The dominant mixing tidal wave component of the circulation can be studied only with a landed network of meteorology stations. An ideal configuration for these would be ~1,000 km apart and encircling the entire planet, meaning that at least 15 or 20 surface stations will be necessary to achieve characterization of the general circulation and seasonal cycle changes in the Mars lower atmosphere (Appendix B: [4.3, 4.5, 4.7]). This need continues to be unmet.

Also unknown is the near-surface moisture profile and the interaction of water vapor with the surface (diurnal diffusive layer). Measurements are needed to understand volatile storage and transport. Multiple in situ humidity stations, configured to measure the height and depth of the "breathing" interface, would be required at optimally chosen locations around the planet (see Chapter 12 in this report).

NEAR-TERM OPPORTUNITIES

The successful launch and entry into orbit of Mars Odyssey in 2001 present the opportunity to make the identifications of surface materials essential to understanding the evolution of the Mars atmosphere. The Gamma-Ray Spectrometer and two supplemental neutron spectrometers will measure major elements and hydrogen globally, thus strongly constraining the surface composition and near-surface water abundances. THEMIS will measure infrared spectra of surface rocks and minerals with 100-m resolution, permitting much better looks through the ubiquitous basalt dust layer, and perhaps identifying compositions.

Plans for the 2003 Mars Exploration Rover missions unfortunately do not include an atmospheric package on the landing stage, analogous to the meteorology station on Mars Pathfinder, a more modest mission.

The European Space Agency's Mars Express mission, planned for launch in 2003, will measure the global distribution of water vapor, ozone, nitrogen, and carbon monoxide with its Planetary Fourier Spectrometer. The Ultraviolet and Infrared Atmospheric Spectrometer (called SPICAM) will use very-high-resolving-power spectroscopy to measure minor atmospheric constituents, including H_2O_2, and C, O, and H isotopic ratios, providing data that will permit the understanding of temporal and spatial changes in atmospheric chemistry. The United Kingdom's Mars Express lander, Beagle 2, will place a package of environmental sensors on the surface to measure ultraviolet radiance, H_2O_2 concentration, gas pressure, and temperature. A gas-analysis package will identify gases released from soil samples and the martian atmosphere, and a mass spectrometer will measure isotopic compositions.

Japan's Nozomi, launched in 1998 and scheduled to reach Mars in 2004, is largely dedicated to studies of Mars's upper atmosphere and the space environment. It will measure upper-atmosphere loss rates, enabling better quantification of the lower atmosphere's odd hydrogen catalytic cycle and the water loss rate.

NASA's Mars Reconnaissance Orbiter, scheduled for launch in 2005, will carry a Pressure-Modulator Infrared Radiometer (PMIRR-MkII) that will globally measure a water vapor band from orbit for 1 full martian year, mapping water vapor, dust, and temperature. MARCI-WA, a wide-angle multichannel ultraviolet and visible imager, will be used to globally and quantitatively map atmospheric O_3, clouds, and hazes.

Earth-Based Observations

Earth-based observations will also continue to play a role in understanding the Mars atmosphere. Especially important will be high-resolving-power spectrographs designed to measure isotopes, especially in the ultraviolet and infrared, from instruments aboard the Stratospheric Observatory for Infrared Astronomy (SOFIA)—a Boeing-747 aircraft modified to carry a 2.4-m telescope. In addition, long-term monitoring of atmospheric water vapor and dust storms can and should be made from smaller, ground-based telescopes. An example of the flexibility of Earth-based measurements is shown in Figure 8.1, where atmospheric water vapor abundances around the Mars Pathfinder landing site are shown in precipitable microns. The ground-based measurements measured ~20 precipitable microns in the atmosphere in the geographical region surrounding the Pathfinder site, somewhat higher than the range of measurements made by the Imager for Mars Pathfinder, 6±4 precipitable microns. The ground-based measurements are important for understanding meteorological conditions in the vicinity of the lander.

Until the surface composition has been thoroughly studied from orbit or in situ, Earth-based measurements will continue to have a role in Mars exploration, as larger telescopes and better instruments now permit the search for spectral features associated with sulfates, carbonates, and other minerals on the martian surface.

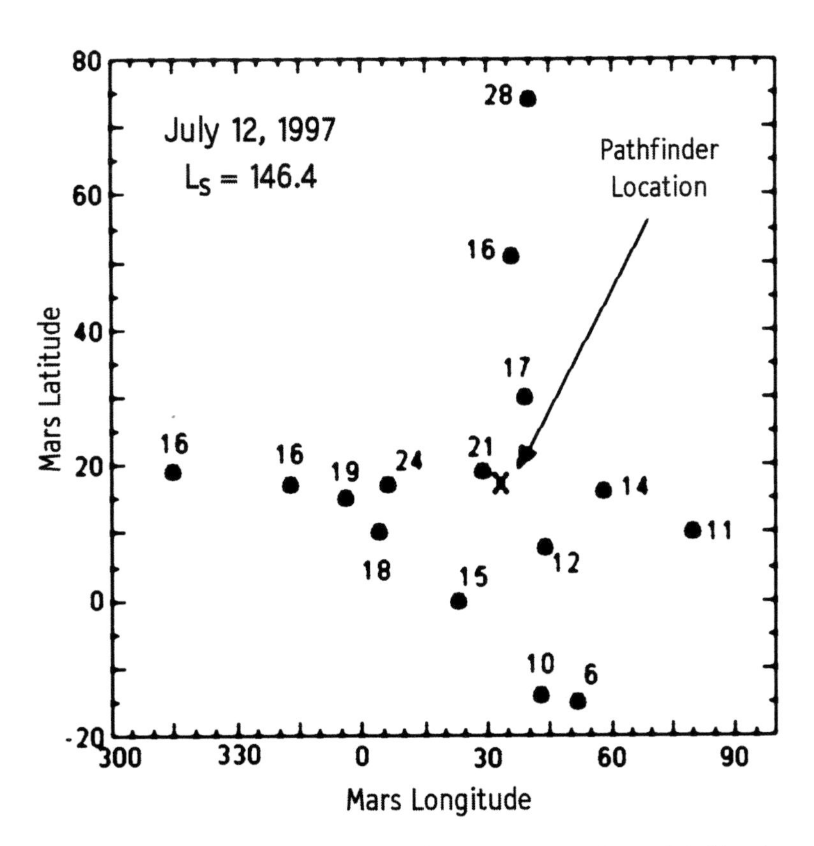

FIGURE 8.1 Earth-based water vapor measurements show column water vapor in precipitable microns at and surrounding the Mars Pathfinder landing site for July 12, 1997, 4 days before in situ measurements by the Imager for Mars Pathfinder experiment found 6±4 precipitable microns (as described in D.V. Titov, W.J. Markiewicz, N. Thomas, H.U. Keller, R.M. Sablotny, M.G. Tomasko, M.T. Lemmon, and P.H. Smith, "Measurements of the Atmospheric Water Vapor on Mars by the Imager for Mars Pathfinder," *Journal of Geophysical Research* 104: 9019–9026, 1999). SOURCE: A.L. Sprague, D.M. Hunten, R.E. Hill, L.R. Doose, and B. Rizk, "Martian Atmospheric Water Abundances: 1996–1999," *Bulletin of the American Astronomical Society* 32: 1093, 2000.

RECOMMENDED SCIENTIFIC PRIORITIES

The martian atmosphere presents questions of meteorology, atmospheric origin and evolution, chemical stability, and atmospheric dynamics. These questions are of particular interest for a broad community, because scientifically useful comparisons with Earth are possible and may prove important for understanding atmospheric evolution not just at Mars but on Earth.

General circulation—the means by which heat, carbon dioxide, water vapor, and dust are transported—is broadly known. Global climate/circulation model simulations have shown that the martian seasonal surface-pressure variation, measured by the Viking landers, has two comparable components—one due to seasonal exchange with the polar caps and the other caused by redistribution of atmospheric mass by the large-scale circulation. Orbiter and lander measurements should be conducted simultaneously, permitting construction of the full three-dimensional circulation. Particular emphasis should be placed on long-term monitoring, with good spatial resolution, of dynamical behavior and changes in local humidity (Appendix B: [1.10, 4.3, 4.5, 4.7, 8.1]). A thorough digestion of the data returned from MGS's TES will go a long way toward improving understanding, but until circulation is measured using Mars-based stations with lifetimes exceeding 1 martian year and spaced between high latitudes and tropical regions, serious ambiguities and gaps in knowledge will remain. The meteorological stations ideally will extend from pole to pole with good meridional spacing of roughly 1,000 km (to sample the baroclinic wave pattern), and each station should measure pressure, temperature, relative humidity, atmospheric opacity, and wind velocity. The European NetLander mission planned for 2007, if successful, will place four meteorological stations on Mars. This is a good beginning for making the type of observations necessary for predictive atmospheric modeling.

As long ago as 1978 (Appendix B: [1]), COMPLEX recognized the need to measure the distribution and abundance of H_2O, CO_2, SO_3, and NO_2 in the martian regolith, and recommended that these volatile compounds be determined to a depth of 2 m with an accuracy of 10 percent of the concentration and a sensitivity of detection of 0.1 percent [1.6]. Also recommended was a complete chemical analysis of the surface material, including all the principal chemical elements (those present in amounts greater than 0.5 percent by atom) as well as those of special biological significance (C, N, Na, P, S, Cl) with a sensitivity of at least 100 ppm [1.7]. The recent MEPAG report [11.2.4, 11.6.2] augments these recommendations by reiterating earlier recommendations to determine the stable isotopic and noble gas composition of the present-day bulk atmosphere.[27] The MEPAG report also recognizes the need for laboratory support to determine the production and reaction rates of key photochemical species (e.g., O_3, H_2O_2, CO, OH) and their interaction with surface materials. All of these issues are of critical importance and were called out in the 1994 COMPLEX report *An Integrated Strategy for the Planetary Sciences: 1995–2010* (Appendix B: 4).[28]

ASSESSMENT OF PRIORITIES IN THE MARS EXPLORATION PROGRAM

In the past, and in this study, COMPLEX has attached very high priority to better understanding the martian atmospheric composition, chemistry, circulation, and concentration of near-surface water vapor as the key components of climate systems and to comparative studies of atmospheric dynamics and evolution. Identification of all atmospheric components present to as low as 10 ppm is essential for the knowledge in its own right and as a baseline for a wide range of other surface-composition and life-detection experiments. Lacking in the current Mars Exploration Program are plans for measurements of the chemical dynamics of C, H, and O by a high-precision, long-lived chemical and isotopic atmospheric analysis at Mars's surface. Time variability of isotopic compositions can be interpreted in terms of sources, sinks, and reservoirs of volatiles, and atmospheric evolution. Isotopic measurements to 3 parts in 10^4 for ^{13}C and ^{18}O, and 3 parts in 10^3 for ^2H are needed to identify dynamic exchange in the current epoch (see Chapter 12 in this report). The current diurnal and seasonal changes in isotopic abundance are essential information for inference of Mars's historical evolution.

NASA's increased emphasis on the search for life on Mars has displaced plans for a meteorology network and studies of atmospheric chemistry and isotope measurements that were given high priority in previous studies. Neither the "Space Science Enterprise Roadmap" section of the *NASA Strategic Plan 2000* (Appendix B: [10]) nor the new NASA Mars Exploration Program (see Appendix A in this report) explicitly addresses these scientific objectives.[29] Under the current NASA Mars Exploration Program, the only avenue for surface meteorology experiments is through the Mars Scout program, with the first such mission scheduled for launch in 2007. The only atmospheric chemistry and isotope measurements projected are those of other nations.

If all investigations planned for launch (including those of the European Space Agency and Japan), as briefly outlined above, are successful and meet their mission objectives, some much-needed measurements will be made. Beagle 2 will search for H_2O_2 and CH_4 at one or a few locations. Atmospheric instruments on Mars Reconnaissance Orbiter will make line-of-sight measurements of water vapor, dust, aerosols, and O_3. Still lacking will be upward-looking spectroscopic measurements from the surface of the planet, which can record diurnal changes in major and minor constituents—CO, O_2, O_3, odd hydrogen, H_2O_2, and water vapor; and the multiyear set of water vapor, wind, and other meteorological measurements that are required to understand Mars's current and past atmosphere.

REFERENCES

1. See, for example, L.A. Leshin, "Insights into Martian Water Reservoirs from Analysis of Martian Meteorite QUE94201," *Geophysical Research Letters* 27: 2017–2020, 2000.
2. P.R. Christensen, J.L. Banfield, M.D. Smith, V.E. Hamilton, and R.N. Clark, "Identification of a Basaltic Component on the Martian Surface from Thermal Emission Spectrometer Data," *Journal of Geophysical Research* 105: 9609–9622, 2000.
3. P.R. Christensen, J.L.Banfield, R.N. Clark, K.S. Edgett, V.E. Hamilton, T. Hoefen, H.H. Keiffer, R.O. Kuzmin, M.D. Lane, M.C. Malin, R.V. Morris, J.C. Pearl, R. Pearson, T.L. Roush, S.W. Ruff, and M.D. Smith, "Detection of Crystalline Hematite Mineralization on Mars by the Thermal Emission Spectrometer: Evidence for Near-Surface Water," *Journal of Geophysical Research* 105: 9623–9642, 2000.
4. A.O. Nier and M.B. McElroy, "Composition and Structure of Mars's Upper Atmosphere: Results from the Neutral Mass Spectrometers on Viking 1 and 2," *Journal of Geophysical Research* 82: 4341–4349, 1977.
5. G.L. Bjoraker, M.J. Mumma, and H.P. Larson, "Isotopic Abundance Ratios for Hydrogen and Oxygen in the Martian Atmosphere," *Bulletin of the American Astronomical Society* 21: 990, 1989.
6. See, for example, H.Y. McSween, "What We Have Learned About Mars from the SNC Meteorites," *Meteoritics* 29: 757–779, 1994.
7. H.J. Melosh and A.M. Vickery, "Impact Erosion of the Primordial Atmosphere of Mars," *Nature* 338: 487–489, 1989.
8. See, for example, J.G. Luhman, R. Johanson, and M.H. Zhang, "Evolutionary Impact of Sputtering of the Martian Atmosphere by O^+ Pick-Up Ions," *Geophysical Research Letters* 19: 2151–2154, 1996.
9. M.B. McElroy and T.M. Donahue, "Stability of the Martian Atmosphere," *Science* 177: 986–988, 1972.
10. T.D. Parkinson and D.M. Hunten, "Spectroscopy and Aeronomy of O_2 on Mars," *Journal of Atmospheric Science* 29: 1380–1390, 1972.
11. T.D. Parkinson and D.M. Hunten, "Spectroscopy and Aeronomy of O_2 on Mars," *Journal of Atmospheric Science* 29: 1380–1390, 1972.
12. B.M. Jakosky and C.B. Farmer, "The Seasonal and Global Behavior of Water Vapor in the Mars Atmosphere: Complete Global Results of the Viking Atmospheric Water Detector Experiment," *Journal of Geophysical Research* 87: 2999–3019, 1982.
13. E.S. Barker, "Martian Atmospheric Water Vapor Observations: 1972-74 Apparition," *Icarus* 28: 247–268, 1976.
14. B. Rizk, W.K. Wells, D.M. Hunten, C.R. Stoker, R.S. Freedman, T. Roush, J.B. Pollack, and R.M. Haberle, "Meridional Martian Water Abundance Profiles During the 1988-1989 Season," *Icarus* 90: 205–213, 1991.
15. A.L. Sprague, D.M. Hunten, R.E. Hill, B. Rizk, and W.K. Wells, "Martian Water Vapor: 1988-1995," *Journal of Geophysical Research* 101: 23229–23241, 1996.
16. D.M. Hunten, A.L. Sprague, and L.R. Doose, "Correction for Dust Opacity of Martian Atmospheric Water Vapor Abundances," *Icarus* 147: 42–48, 2000.
17. C. Barth, C.W. Hord, A.I. Stewart, A.L. Lane, M.L. Dick, and G.P. Anderson, "Mariner 9 Ultraviolet Spectrometer Experiment: Seasonal Variation of Ozone on Mars," *Science* 179: 795–796, 1973.
18. See, for example, C. Leovy, R. Zurek, and J. Pollack, "Mechanisms of Mars Dust Storms," *Journal of Atmospheric Science* 30: 749–762, 1973.
19. M.G. Tomasko, L.R. Doose, M. Lemmon, P.H. Smith, and E. Wegryn, "Properties of Dust in the Martian Atmosphere from the Imager for Mars Pathfinder," *Journal of Geophysical Research* 104: 8987–9007, 1999.
20. B.A. Cantor, P.B. James, M. Caplinger, M.C. Malin, and K.S. Edgett, "Martian Dust Storms: 1999 MOC Observations," p. 20 in *Second International Conference on Mars Polar Science and Exploration, August 21–25, 2000, Reykjavik, Iceland*, LPI Contribution No. 1057, Lunar and Planetary Institute, Houston, Texas, 2000.

21. M.E. Ockert-Bell, J. Bell III, J. Pollack, C.P. McKay, and F. Forget, "Absorption and Scattering Properties of the Mars Dust in Solar Wavelengths," *Journal of Geophysical Research* 102: 9039–9050, 1997.
22. N. Thomas, W.J. Markiewicz, R.M. Sablotny, M.W. Wuttke, H.U. Keller, J.R. Johnson, R.J. Reid, and P.H. Smith, "The Color of the Martian Sky and its Influence on the Illumination of the Martian Surface," *Journal of Geophysical Research* 104: 8795–8808, 1999.
23. B.J. Conrath, J.C. Pearl, M.D. Smith, W.C. Maguire, P.R. Christensen, S. Dason, and M.S. Kaelberer, "Mars Global Surveyor Thermal Emission Spectrometer (TES) Observations: Atmospheric Temperatures During Aerobraking and Science Phasing," *Journal of Geophysical Research* 105: 9509–9520, 2000.
24. D. Bandfield, B. Conrath, J.C. Pearl, M.D. Smith, and P. Christensen, "Thermal Tides and Stationary Waves on Mars as Revealed by Mars Global Surveyor Thermal Emission Spectrometer," *Journal of Geophysical Research* 105: 9521–9538, 2000.
25. A.J. Kliore, D.L. Cain, G. Fjeldbo, B.L. Seidel, M.J. Sykes, and S.I. Rasool, "The Atmosphere of Mars from Mariner 9 Radio Occultation Measurements," *Icarus* 17: 484–516, 1972.
26. R.M. Haberle, M.M. Joshi, J.R. Murphy, J.R. Barnes, J.T. Schofield, G. Wilson, M. Lopez-Valverde, J.L. Hollingsworth, A.F. Bridger, and J. Schaeffer, "General Circulation Model Simulations of the Mars Pathfinder Atmospheric Structure Investigation/Meteorology Data," *Journal of Geophysical Research* 104: 8957–8974, 1999.
27. NASA, Mars Exploration Payload Assessment Group (MEPAG), "Mars Exploration Program: Scientific Goals, Objectives, Investigations, and Priorities," December 2000, in *Science Planning for Exploring Mars*, JPL Publication 01-7, Jet Propulsion Laboratory, Pasadena, Calif., 2001.
28. Space Studies Board, National Research Council, *An Integrated Strategy for the Planetary Sciences: 1995–2010*, National Academy Press, Washington, D.C., 1994.
29. National Aeronautics and Space Administration, *Strategic Plan 2000*, NASA, Washington, D.C., 2000.

9

Climate Change

PRESENT STATE OF KNOWLEDGE

Climate encompasses a broad range of complex, interacting systems with a wide range of time scales. The Mars climate system, which includes the surface, atmosphere, polar caps, and accessible regions of the subsurface, has undergone significant change during the planet's history. Our knowledge of the martian climate can be organized according to the three time scales over which the climate is expected to vary: (1) interannual climatic variability, (2) quasi-periodic climate variations, and (3) long-term climate change.

Interannual Climatic Variability

Martian climate variability on interannual time scales is an interesting topic from many points of view. Because of the short radiative time constant of the martian atmosphere and its lack of ocean heat storage, one would not expect the martian climate to undergo significant variations on interannual time scales. Yet, with detailed examination of the multidecade telescopic record of great dust storms,[1] multiyear surface pressure records acquired at the Viking landing sites,[2] multiyear orbiter observations of the appearance of the seasonal and residual polar caps,[3] and evidence of large variations in atmospheric water, it becomes clear that the climate of Mars does exhibit distinct variations from one year to the next. Understanding the nature and causes of these variations is important for identifying interactions among the cycles of carbon dioxide, dust, and water in Mars's present climate. Understanding the nature and causes of these variations is also important for the study of longer-term climate variations, which can potentially be thought of as the cumulative effects of interannual variability. The incomplete set of observations of martian interannual variations that exists today exhibits no clear cycles, trends, or patterns, and it may be consistent with the hypothesis that these variations result from random drift.[4]

Quasi-Periodic Climate Variations

One of the cornerstones of our understanding of Earth's climate is that small quasi-periodic variations in Earth's orbital and axial elements over time scales of tens to hundreds of thousands of years result in large-scale changes in this planet's climate.[5] The effects of these quasi-periodic climate variations are strongest at high latitudes, which have experienced repeated cycles of glaciation for billions of years.[6] Mars's orbital and axial elements experience variability on time scales that are comparable with those of Earth, but the magnitudes of these variations for Mars are significantly greater.[7] For instance, orbital calculations show that the obliquity of the martian spin axis has deviated from its present value of 25.1° to become as high as 35° within the past million years, and as high as 47° for brief periods within the past 10 million years.[8] The consequent changes to the insolation at high latitudes (see Figure 9.1) undoubtedly have caused significant changes in the seasonal cycles of carbon dioxide, water, and dust.[9] On the basis of present understanding, Mars is the planet in the solar system that one would expect to experience the most significant quasi-periodic variations in its climate.

On Earth, physical and chemical climate records have proved to be the most valuable tools for studying and understanding quasi-periodic climate variability. Ice cores obtained in Greenland and Antarctica and deep-sea cores of ocean sediments contain a wealth of nearly continuous information about the state of the global climate system over the past 200,000 years.[10] The exploration of Mars is at a much earlier stage than that of Earth, but what has been observed thus far suggests very strongly that accessible records of past martian climate variations may exist in the form of layered deposits in the martian polar regions[11] and at midlatitudes.[12]

Polar layered deposits visible on exposed scarps were first identified in Mariner 9 images,[13] and subsequent observations by Viking and Mars Global Surveyor (MGS) have revealed extensive layering within both the north and the south polar caps and their surrounding regions (see Figure 9.2). In recent images from the Mars Orbiter Camera (MOC) on MGS, individual layers can be traced for hundreds of kilometers, and in some locations the vertical scale of the layered structure extends down to the 1.5-m resolution limit of the camera.[14] Fine-scale layered deposits, which are most likely of sedimentary origin, have recently been observed by MOC at low latitudes within impact craters and canyons.[15]

Despite the excitement that the orbiter observations of layering on Mars have generated, the limitations of the orbital perspective have made it difficult to learn anything definitive about the composition, origin, or age of the layers observed in the images. Clearly, more knowledge is needed before one can conclude that these layers are true climatological records and that they can be deciphered. Only then can researchers make connections between

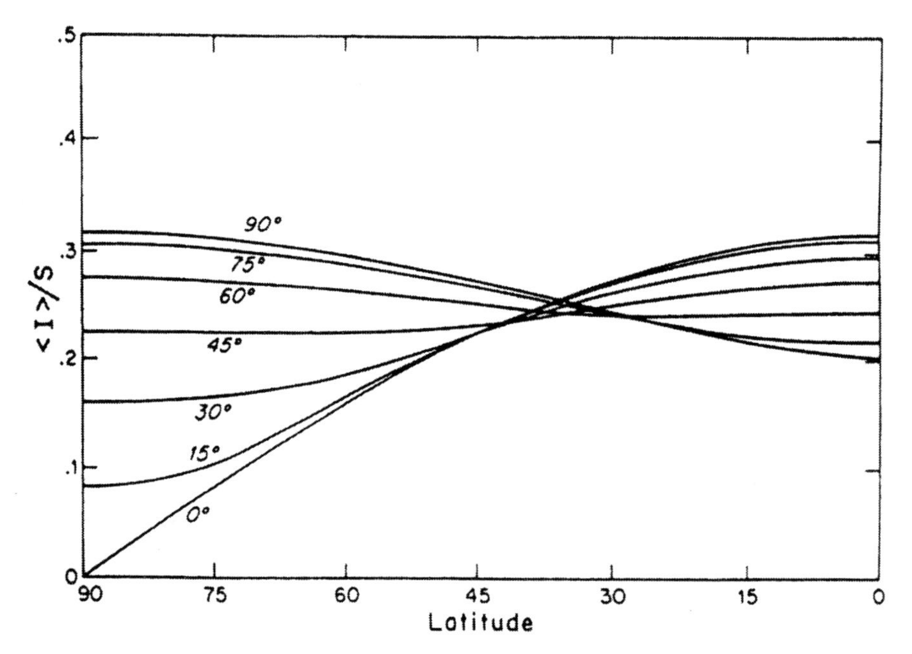

FIGURE 9.1 Average annual insolation < I >, normalized to the average solar flux S, as a function of latitude for various values of the obliquity. At obliquities higher than 54° the average annual insolation is higher at the poles than at the equator. SOURCE: W.R. Ward, "Long-Term Orbital and Spin Dynamics of Mars," pp. 298–320 in *Mars*, H.H. Kieffer, B.M. Jakosky, C.W. Snyder, and M.S. Matthews (eds.), University of Arizona Press, Tucson, 1992. Copyright 1992 by the Arizona Board of Regents. Reprinted by permission of the University of Arizona Press.

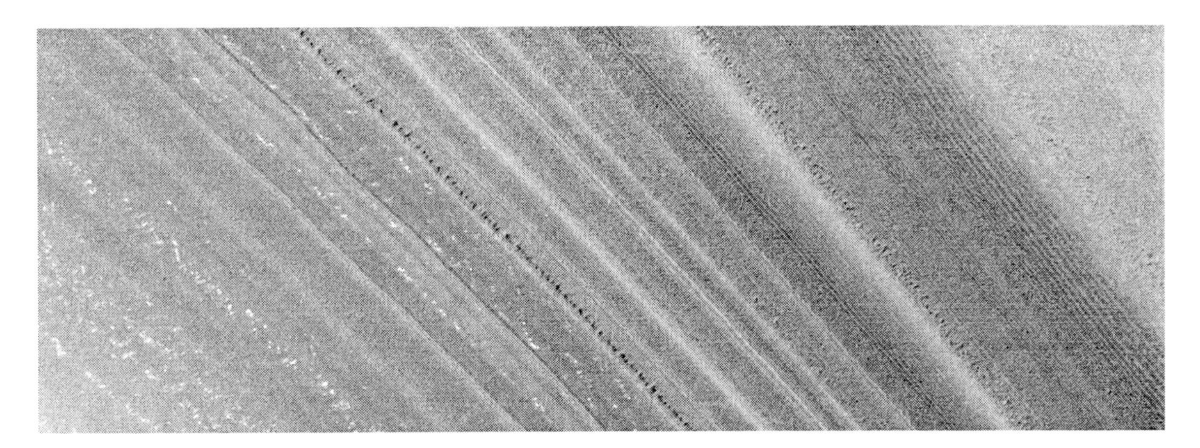

FIGURE 9.2 Mars Global Surveyor MOC image of fine-scale layering in the martian north polar water ice cap. The image is centered at 87.0°N, 263.8°W, and is approximately 2 km × 4 km in size. SOURCE: MGS MOC Release No. MOC2-266, December 22, 2000. Courtesy of NASA/JPL/Malin Space Science Systems.

the external and internal forcing factors that have driven the martian climate, and the response of the martian climate system through time.

Long-Term Climate Change

The idea that Mars was once warm and wet, though it is now cold and dry, dates back to before the time of Percival Lowell.[16] Today, the evidence suggests that this speculation was correct, in that there is a wide range of surface features on the planet that can be interpreted as evidence for warmer climatic conditions at various times in Mars's history. However, considerable uncertainty remains as to whether these features uniquely point to the existence of a stable, warm, wet climate, or whether they were formed in episodic nonequilibrium events.[17]

From the standpoint of theoretical climate studies, there is general consensus that Mars possesses all the volatile ingredients necessary to produce a warm and wet climate.[18] The problem is that at Mars's distance from the Sun, the stable location for Mars's volatiles is not in the atmosphere but in condensed phases, which makes it difficult to maintain a stable martian greenhouse (see Figure 9.3). This problem is compounded by the fact that solar evolution models predict that the solar constant was significantly lower billions of years ago.[19] Nonetheless, the extreme complexity of the climate system and the many interactions between its various components and external forcing factors, such as solar variability, orbital variability, volcanism, and meteoritic and cometary impacts, make it possible that the early climate of Mars was warm and wet.

Recent efforts to understand the history of volatiles and climate on Mars have established a broad framework that is consistent with evidence from spacecraft data, laboratory investigations, and theoretical studies.[20] Although the earliest atmosphere was probably lost by impact erosion and hydrodynamic escape during the Early Noachian epoch, a relatively robust atmosphere appears to have been reestablished during the Noachian by primitive volatiles released during the creation of Tharsis by volcanic and igneous processes. These volatiles were likely partially sequestered into reservoirs such as carbonates or lost to space, although the relative amounts are unknown. The end of the Noachian epoch marked a huge change in the climate and probably the volatile inventory of Mars: erosion rates declined, valley network formation largely ceased, and magmatism declined along with the decline or even cessation of the intrinsic magnetic field. The loss of the protective magnetic field likely resulted in a substantial erosion of the atmosphere. As discussed in Chapters 5 and 6 in this report, this did not mark the complete arrest of important changes in the climate or volatile inventories (e.g., large outflow channels, continued redistribution of volatiles). However, it did mark a fundamental change of state of the atmosphere. While this provides a framework for understanding the martian climate and volatile history consistent with available data, there are still many aspects of these hypotheses that need to be tested.

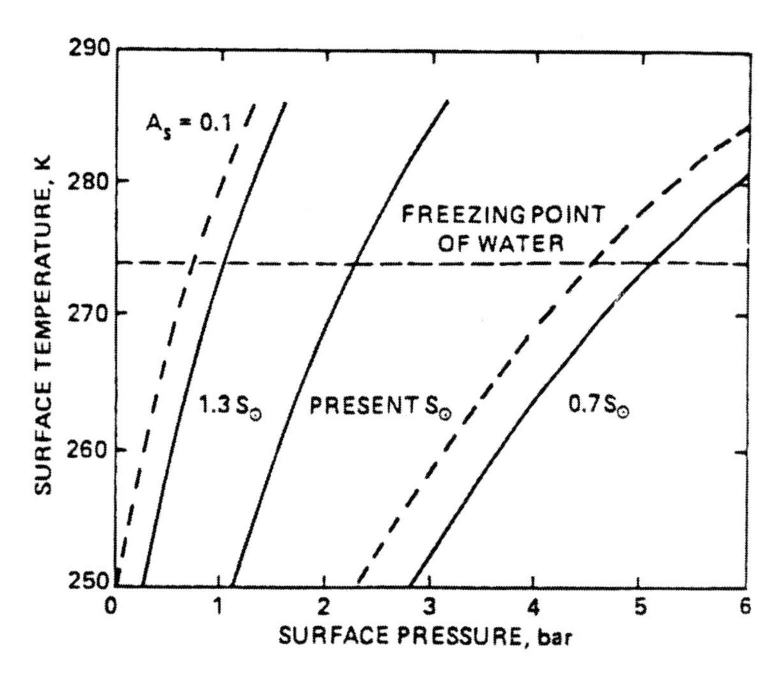

FIGURE 9.3 Surface temperature on Mars as a function of the surface pressure. The model is a one-dimensional, globally averaged, radiative equilibrium model with a convective troposphere. The surface pressure fixes the optical depth of CO_2; water is assumed to be 77 percent saturated at the surface with a relative humidity decreasing with height. This defines water vapor optical properties in terms of temperature, so that the solution must be iterated. The solid lines are for a surface albedo with the present value (0.215); the broken lines are for an albedo of 0.1. Calculations are made for three values of the solar flux; 0.7 times the present flux ($S_\odot$) corresponds to conditions expected in the Noachian epoch. The pressure of atmospheric gases for liquid water to exist on the surface varies from 2.2 atm today to 5 atm in the Noachian. SOURCE: Reprinted from J.B. Pollack, J.F. Kasting, S.M. Richardson, and K. Poliakoff, "The Case for a Warm Wet Climate on Early Mars," *Icarus* 71: 203–224, 1987, copyright 1987, with permission from Elsevier.

As with the subject of quasi-periodic climate change, the study of long-term climate change on Mars is in desperate need of some interpretable climatologic evidence. Such evidence could come from the discovery of extensive water-laid sediments, or extensive deposits of carbonates, hydrates, or evaporites. If this evidence could be coupled with a better understanding of martian chronology tied to absolute dating, so much the better. In summary, there is every reason to expect that Mars has had a rich and variable climatic history, but to date observations and theory have given only tantalizing glimpses of that history.

NEAR-TERM OPPORTUNITIES

The failed Mars Polar Lander mission, which attempted to land in the south polar layered deposits in 1999, represented a unique opportunity to learn more about the nature and potential of the deposits as records of Mars's past climatic history.

The Mars Odyssey orbiter, successfully launched in April 2001 and now operating in orbit about Mars, includes the Gamma-Ray Spectrometer (GRS) instrument that was originally flown on the failed Mars Observer orbiter in 1993. The GRS will produce the first map of the global elemental composition of the martian surface. Geographic variations in composition may provide clues to the locations of extensive sedimentary deposits or other clues to Mars's past climate.

Looking farther into the future, the Mars Exploration Rovers to be launched in 2003 will be equipped with instrumentation designed to learn more about the mineralogic and elemental composition of martian rocks and soils, which could yield information relevant to Mars's past climate. Mars Odyssey's neutron spectrometer is measuring (with coarse resolution) the near-surface water abundance in the polar layered deposits and residual water-ice cap, the water abundance in the south-polar residual cap, and the seasonal cycle of CO_2 frost via the determination of spatially and temporally resolved frost thickness. The Mars Reconnaissance Orbiter, scheduled for launch in 2005, will include a high-resolution imager capable of acquiring images of layered deposits at a resolution of 0.6 m, and the atmospheric sounder PMIRR-Mk II. However, on the whole, few planned missions or experiments beyond the current suite of missions are focused on climate-related objectives.

RECOMMENDED SCIENTIFIC PRIORITIES

Understanding the climate of Mars involves a broad range of measurements and scientific priorities, and climate has become increasingly prominent in recommendations regarding scientific priorities for Mars exploration. Early recommendations focused on understanding Mars's volatile inventory and interactions between its volatile cycles, as well as understanding its geological and geochemical history (Appendix B: [1.10, 4.4]). Later reports called attention to the need to better understand the martian atmospheric circulation as a key component of the planet's climate system [4.3, 4.5, 4.7, 8.1]. In the Mars Exploration Payload Assessment Group (MEPAG) report, climate is explicitly called out as one of four organizing themes for Mars scientific exploration.[21] High-priority climate-related investigations recommended in the MEPAG report include determining the processes controlling the present distributions of Mars volatiles and dust using global mapping and landed observations [11.3.3], and finding physical and chemical records of past climates using remote sensing, landed exploration, and returned samples [11.1.5, 11.2.5].

ASSESSMENT OF PRIORITIES IN THE MARS EXPLORATION PROGRAM

There is little in the Mars Exploration Program that explicitly addresses questions of climate. Landers and rovers should be sent to the polar regions to better define near-surface volatile properties and behavior and to allow better understanding of the nature of polar layered deposits and the climate records they may contain; this was the goal of the failed Mars Polar Lander mission. However, it should be understood that finding definitive records of past climates is difficult, even on Earth. NASA's strategy of surface geological exploration of promising sites, combined with detailed analyses of returned samples, may lead in time to important conclusions regarding the Mars climate, but clear-cut results will require the exploration of a large number of widely distributed sites and the detailed analysis of many samples.

REFERENCES

1. R.W. Zurek and L.J. Martin, "Interannual Variability of Planet-Encircling Dust Storms on Mars," *Journal of Geophysical Research* 98: 3247–3259, 1993.
2. R.W. Zurek, J.R. Barnes, R.M. Haberle, J.B. Pollack, J.E. Tillman, and C.B. Leovy, "Dynamics of the Atmosphere of Mars," pp. 835–933 in *Mars*, H.H. Kieffer, B.M. Jakosky, C.W. Snyder, and M.S. Matthews (eds.), University of Arizona Press, Tucson, 1992.
3. P.B. James, H.H. Kieffer, and D.A. Paige, "The Seasonal Cycle of Carbon Dioxide on Mars," pp. 934–968 in *Mars*, H.H. Kieffer, B.M. Jakosky, C.W. Snyder, and M.S. Matthews (eds.), University of Arizona Press, Tucson, 1992.
4. A.P. Ingersoll, and J.R. Lyons, "Mars Dust Storms: Interannual Variability and Chaos," *Journal of Geophysical Research* 98: 10951–10961, 1993.
5. J. Imbrie, "Astronomical Theory of the Pleistocene Ice Ages: A Brief Historical Review," *Icarus* 50: 408–422, 1982.
6. N. Christie-Blick, I.A. Dyson, and C.C. von der Borch, "Sequence Stratigraphy and the Interpretation of Neoproterozoic Earth History," *Precambrian Research* 73: 3–26, 1995.
7. W.R. Ward, "Long-Term Orbital and Spin Dynamics of Mars," pp. 298–320 in *Mars*, H.H. Kieffer, B.M. Jakosky, C.W. Snyder, and M.S. Matthews (eds.), University of Arizona Press, Tucson, 1992.
8. J. Touma and J. Wisdom, "The Chaotic Obliquity of Mars," *Science* 259: 1294–1297, 1993.

9. O.B. Toon, J.B. Pollack, W. Ward, J.A. Burns, and K. Bilski, "The Astronomical Theory of Climatic Change on Mars," *Icarus* 44: 552–607, 1980.
10. J.R. Petit, I. Basile, A. Leruyuet, D. Raynaud, C. Lorius, J. Jouzel, M. Stievenard, V.Y. Lipenkov, N.I. Barkov, B.B. Kudryashov, M. Davis, E. Saltzman, and V. Kotlyakov, "Four Climate Cycles in Vostok Ice Core," *Nature* 387: 359–360, 1997.
11. P. Thomas, S. Squyres, K. Herkenhoff, A. Howard, and B. Murray, "Polar Deposits on Mars," pp. 767–798 in *Mars*, H.H. Kieffer, B.M. Jakosky, C.W. Snyder, and M.S. Matthews (eds.), University of Arizona Press, Tucson, 1992.
12. M.C. Malin and K.S. Edgett, "Sedimentary Rocks of Early Mars," *Science* 290: 1927–1937, 2000.
13. B.C. Murray, L.A. Soderblom, J.A. Cutts, R.P. Sharp, D.J. Milton, and R.B. Leighton, "Geological Framework for the South Polar Region of Mars," *Icarus* 17: 328–345, 1972.
14. M.C. Malin, M.H. Carr, G.E. Danielson, M.E. Davies, W.K. Hartmann, A.P. Ingersoll, P.B. James, H. Masursky, A.S. McEwen, L.A. Soderblom, P. Thomas, J. Veverka, M.A. Caplinger, M.A. Ravine, T.A. Soulanille, and J.L. Warren, "Early Views of the Martian Surface from the Mars Orbiter Camera of the Mars Global Surveyor," *Science* 279: 1681–1685, 1998.
15. M.C. Malin and K.S. Edgett, "Sedimentary Rocks of Early Mars," *Science* 290: 1927–1937, 2000.
16. P. Lowell, *Mars As the Abode of Life*, Macmillan, New York, 1908.
17. S.W. Squyres, "Urey Prize Lecture: Water on Mars," *Icarus* 79:229–288, 1989.
18. C.P. McKay, R.L. Mancinelli, C.R. Stoker, and R.A. Wharton, Jr., "The Possibility of Life on Mars During a Water-Rich Past," pp. 1234–1245 in *Mars*, H.H. Kieffer, B.M. Jakosky, C.W. Snyder, and M.S. Matthews (eds.), University of Arizona Press, Tucson, Arizona, 1992.
19. J.F. Kasting, "Update: The Early Mars Climate Question Heats Up," *Science* 278: 1245, 1997.
20. See, for example, the review by B.M. Jakosky, and R.J. Phillips, "Mars' Volatile and Climate History, *Nature* 412: 237–244, 2001.
21. NASA, Mars Exploration Payload Assessment Group (MEPAG), "Mars Exploration Program: Scientific Goals, Objectives, Investigations, and Priorities," December 2000, in *Science Planning for Exploring Mars*, JPL Publication 01-7, Jet Propulsion Laboratory, Pasadena, Calif., 2001.

10

Upper Atmosphere, Ionosphere, and Solar Wind Interaction

PRESENT STATE OF KNOWLEDGE

The Upper Atmosphere

Very little information is available on the upper atmosphere of Mars. The only in situ measurements of atmospheric composition came from neutral mass spectrometers on the two Viking landers in 1976.[1] These provided two midlatitude vertical profiles, in the altitude range of about 120 to 200 km, of CO_2, CO, N_2, O_2, and Ar densities, corresponding to solar zenith angles of approximately 44°, during conditions of low solar activity. Through use of the scale heights thus measured, atmospheric temperature profiles were deduced. These temperatures showed quite large variations, with an average value of less than 200 K.

Some indirect and limited information on composition and temperatures has been obtained using airglow and ionospheric information.[2] The upper-atmospheric temperatures appear to vary by about 150 K between solar cycle minimum and maximum conditions. Estimates of atomic oxygen densities have been obtained from ion density measurements, to be discussed below, and ultraviolet spectroscopy. CO_2 is the major neutral constituent below about 200 km, above which atomic oxygen predominates. However, these conclusions are based on only a couple of measurements made at a particular solar zenith angle and solar cycle condition. Lyman-α airglow observations have provided information on the daytime thermal hydrogen densities, indicating a value on the order of 10^5 cm^{-3} at an altitude of 150 km.[3] The first observations of a hot atom corona anywhere in the solar system were provided by the Lyman-α measurements of hydrogen at Venus.[4] An extended hot oxygen and carbon corona has also been observed at Venus.[5,6] Although no corresponding observations are yet available for Mars, theoretical models predict the presence of a similarly hot atom corona.[7,8]

The z-axis accelerometer carried by the Mars Global Surveyor (MGS) provided a great deal of important information about total densities and temperatures during its extended aerobraking period.[9] It made measurements over about 900 orbits between altitudes of approximately 110 to 160 km during solar minimum to moderate conditions. The observed longitude-fixed density variations in the Mars lower thermosphere are thought to be generated by the modulation of thermal tides by the significant Mars topography. The MGS accelerometer witnessed the onset, rise, and decay of a regional dust storm event, and the corresponding responses of the upper-atmosphere densities throughout this "Noachis" dust storm. A three-fold increase in the 130-km densities was observed at 40°N latitude, approximately two to three orbits (a few days) after the Noachis storm was detected by other instruments. A roughly 8-km expansion of the thermosphere was also seen over a few days at onset, with a subsequent contraction of the thermosphere back to original levels weeks later. Comprehensive thermospheric general circulation models are now in existence (see Figure 10.1), and they are quite successful in accounting for the overall MGS temperature observations.[10] However, the dynamical and radiative processes that drive the Mars lower and upper atmospheres on the short time scales corresponding to short-lived dust storms are yet to be explained, and of course no direct information on the winds is available.

Ionosphere and Solar Wind Interaction

The only in situ measurements of the thermal plasma composition, density, and temperature in the ionosphere of Mars were obtained by the retarding potential analyzers carried aboard the two Viking landers[11] and by the mass spectrometers mentioned above. The retarding potential analyzers provided two vertical profiles of the densities of the three most abundant ions (O_2^+, O^+, and CO_2^+) in the altitude region of about 120 to 300 km. The observations confirmed theoretical suggestions that the most abundant ion is O_2^+; at first this seemed a surprising result because of the practically total absence of neutral molecular oxygen in the upper atmosphere of Mars. This finding demonstrates the importance of ion chemistry in ionospheres. The retarding potential analyzers also provided information on the ion and electron temperatures, but in a limited altitude range and along only two profiles.[12] These temperatures were found to be a few thousand degrees, which cannot be explained by extreme ultraviolet

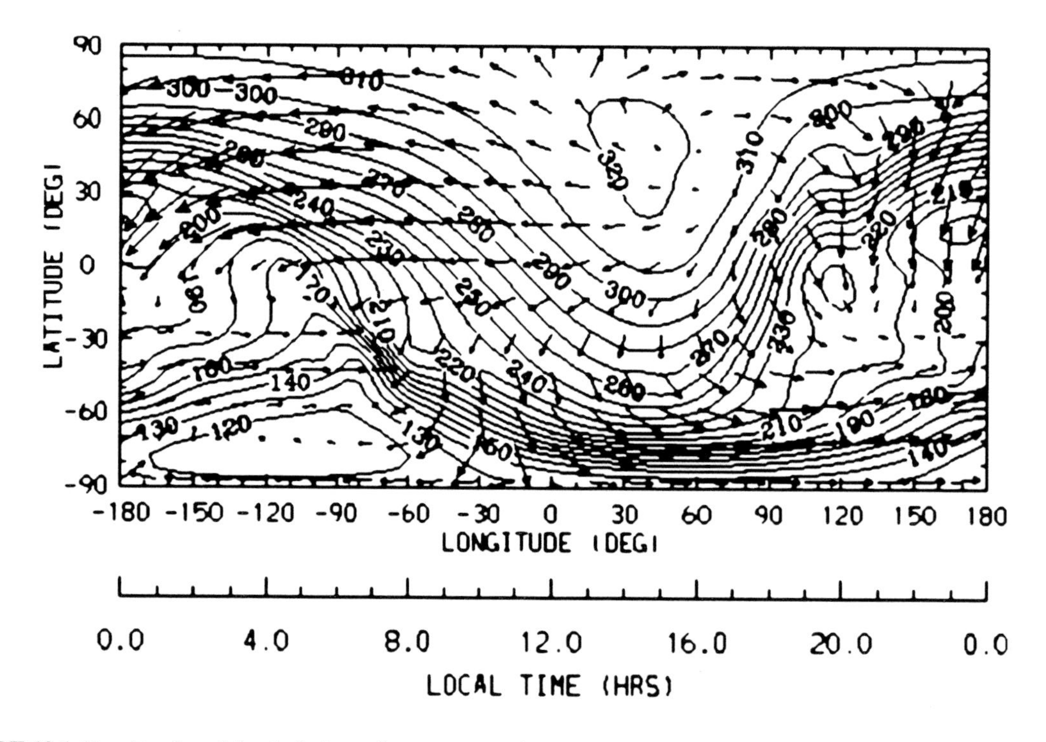

FIGURE 10.1 Results of model calculations of upper-atmosphere temperatures and winds at Mars, in a plot of latitude versus local solar time (LST), for solar maximum and northern summer conditions at ~200-km altitude. The isotherms shown are in 10-K intervals; superimposed arrows represent the magnitude and direction of the neutral winds. The winds diverge from midafternoon and converge after dusk or before dawn. Neutral temperatures reach 321 K (dayside, LST = 16) and decline to 111 K (south polar night). Maximum winds reach 326 m/s across the terminators and near the poles. SOURCE: S.W. Bougher, S. Engel, R.G. Roble, and B. Foster, "Comparative Terrestrial Planet Thermospheres: 3. Solar Cycle Variation of Global Structure and Winds at Solstices," *Journal of Geophysical Research* 105: 17669-17692, 2000. Copyright 2000 by the American Geophysical Union. Reproduced by permission of AGU.

heating and classical thermal conduction alone, as is the case in the terrestrial midlatitude ionosphere. Similar results were seen at Venus, and it is now believed that a combination of reduced thermal conduction and additional topside heat input is the cause of these enhanced temperatures.[13] An example of such a calculation, along with the observed temperatures, is shown in Figure 10.2. The required heat inflow is considerably higher for the electrons, mainly because of their larger thermal conductivity, which leads to rapid drainage of heat to the neutrals at lower altitudes.

Electron density altitude profiles were also obtained by several U.S. and Soviet satellites (e.g., Mariner 9), using the radio occultation technique. Thus, some information exists on both the dayside and near-terminator-nightside electron density values, covering the altitude range of about 120 to 300 km.[14,15] No clear presence of an ionopause was seen in this database.

A number of the U.S. and Soviet spacecraft that either flew by or orbited Mars carried magnetometers and some limited plasma instrumentation; they discovered a well-defined bow shock around the planet and provided limited information on the fields and particles inside the bow shock. The Soviet Phobos mission was the first (and so far the only) spacecraft that has orbited Mars and that had a comprehensive fields and particles instrument complement.[16] Phobos made important additions to our database on the bow shock location[17] and provided new information on extensive wave activities[18] and the presence of a magnetic pile-up boundary (also called a magnetic barrier). This latter phenomenon is the result of the slowing down of the shocked interplanetary magnetic field inside the magnetosheath. The Phobos spacecraft, after only a few elliptic orbits, was placed in a circular orbit at about 3 Mars radii (the orbit of the satellite Phobos), so it provided very little spatial and temporal information on the dayside magnetosheath. Also, the Phobos spacecraft, even during the elliptical phase of its orbit, did not get closer than about 850 km to the surface, and therefore no ionospheric information was obtained. This orbit did not

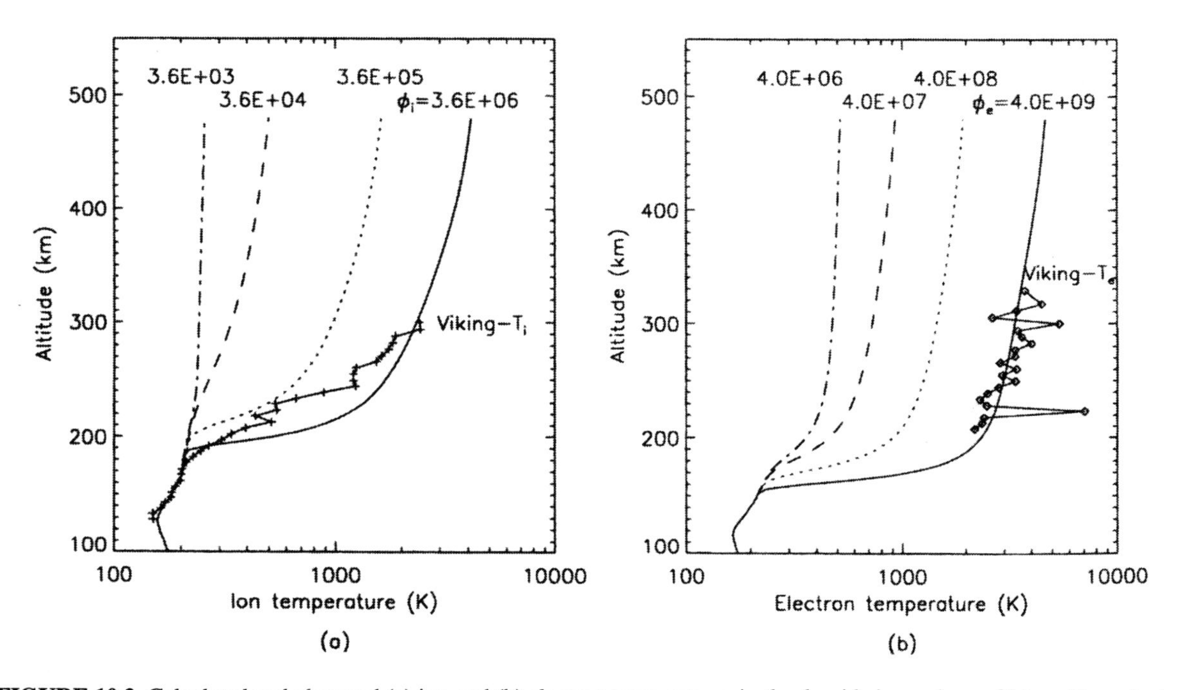

FIGURE 10.2 Calculated and observed (a) ion and (b) electron temperatures in the dayside ionosphere of Mars. The calculations assume different topside heat inflows, as indicated at the tops of the curves. The inflow numbers given are in eV cm^{-2} sec^{-1}. The heat inflow values necessary to match the temperatures observed by Viking are reasonable, but they are ad hoc values. SOURCE: Y.W. Choi, J. Kim, K.W. Min, A.F. Nagy, and K.I. Oyama, "Effect of the Magnetic Field on the Energetics of Mars' Ionosphere," *Geophysical Research Letters* 25: 2753-2756, 1998. Copyright 1998 by the American Geophysical Union. Reproduced by permission of AGU.

allow a definitive conclusion to be drawn about the existence of an intrinsic magnetic field at Mars. These comments should not be construed to imply that Phobos did not provide important new information concerning the plasma environment of Mars—only that much more data will be needed to permit elucidation of the controlling physical processes.

Mars Global Surveyor carried a magnetometer and electron reflectometer package, and during the extended aerobraking phase of the mission it made many low-altitude measurements. One of the most exciting and unexpected results of the MGS mission was the finding that although Mars has no intrinsic magnetic field, relatively strong and widespread remnant crustal magnetization is present (see Chapter 2 in this report). In the context of this section, the lack of an intrinsic magnetic field and the presence of strong, localized crustal magnetic fields have important implications for the interaction of the solar wind with Mars. It means that in most situations, the interaction is with the ionosphere, as is the case for Venus, but asymmetries must be present because of the nonuniform distribution of the crustal fields. The presence of these crustal fields also means that "mini-magnetospheres" are present and some form of intermittent reconnection processes must take place. The MGS magnetometer also provided information on the nature and extent of the magnetic pile-up boundary.[19] The electron reflectometer measured superthermal electron fluxes in both the magnetosheath and the ionosphere. It found distinct changes in the shape of the electron spectrum, in the energy range of 10 to 10,000 eV, in these two regions,[20] and thus it established, unambiguously and for the first time, the presence of an ionopause, which had been expected but not seen before.

NEAR-TERM OPPORTUNITIES

Nozomi

The Japanese mission Nozomi was specially designed and implemented to address many of the outstanding problems in the areas of upper-atmospheric and plasma sciences associated with Mars (see Table A.1 in Appendix A). However, even this dedicated payload does not address all of the high-priority science that needs to be covered; for example there are no measurements of the upper-atmospheric winds, which are crucial for an understanding of the dynamics of that region.

The spacecraft was launched on July 4, 1998, and was supposed to be placed in orbit around Mars in March of the following year. However, because of a sticking valve problem, the planned trajectory had to be abandoned; after two Earth swingbys, it is expected to be placed in orbit around Mars early in 2004. Nozomi was not designed for such a long interplanetary cruise, and it now faces a number of problems. If successful, the mission will answer a number of outstanding questions (see the next two subsections and Table 12.1, "Outstanding Mars Exploration Science Issues," in Chapter 12), but it is clear that even then it cannot and will not "close the book" on this field of research. Besides the mission's lack of certain important measurements, the 5-year delay means that instead of arriving at Mars during solar cycle maximum, it will be placed in orbit during solar cycle minimum, and thus the meaningful investigation of a number of outstanding problems will be seriously jeopardized (e.g., nonthermal escape, which is highly solar-cycle dependent).

Mars Express

The European Space Agency's Mars Express mission is expected to be launched in June 2003, to take 6 months to reach Mars, and to be put in orbit around the planet. Of its instruments, three have relevance to this section (see Table A.1). They are the Ultraviolet and Infrared Atmospheric Spectrometer (SPICAM), the Energetic Neutral Atoms Analyzer instrument package (called ASPERA), and the Radio Science Experiment (known as MaRS). It is hoped that SPICAM will provide some information on the thermal and hot neutral gas distributions. ASPERA is designed to study the solar wind interaction with Mars, but all three of these science packages will also provide some information on the plasma densities around the planet.

RECOMMENDED SCIENTIFIC PRIORITIES

In the general area of upper-atmosphere, ionosphere, and solar wind interaction studies of Mars, the main priorities for research are these:

1. *The dynamics of the upper atmosphere.* Absolutely no direct information exists on neutral gas velocities, but it is badly needed in order to obtain a basic understanding of the dynamical processes and coupling to the lower atmosphere.
2. *Hot atom abundances and escape fluxes.* Pioneer Venus's ultraviolet spectrometer established the presence of hot oxygen and carbon atoms at Venus, and the earlier Mariner observations indicated the presence of hot hydrogen. There is no information on these hot atoms at Mars, but in light of the similarities with Venus and of theoretical models, it is clear that they must be present. The low escape energy at Mars also implies that hot atom escape fluxes may be important.
3. *Ion escape from Mars.* Both theory and observations indicate that there are significant ion escape fluxes at Mars. These escape flows, whether due to tailward flow from the ionosphere or to scavenging, are related to the solar wind interaction processes. A meaningful understanding of the escape processes, mentioned under priorities 2 and 3 above, are of great importance in understanding how the martian atmosphere evolved during the last few billion years. If there was life on Mars some time ago, how did an atmosphere that was capable of supporting life evolve into the one that remains today?
4. *Mini-magnetospheres and reconnection at Mars.* The discovery of remnant crustal magnetic fields at Mars means that small, localized "magnetospheres" are likely to be present. There have been suggestions that such mini-magnetospheres are present around the Moon. These small magnetospheric regions may undergo reconnection with the compressed interplanetary magnetic field in the magnetosheath. These reconnection events must depend on the specific location of the crustal field with respect to the subsolar location and solar wind parameters, such as magnetic field angle.
5. *The energetics of the ionosphere.* There are major unexplained issues concerning the mechanisms that determine the electron and ion temperatures in the ionospheres of both Venus and Mars. The question is why the electron and ion temperatures are as high as the measurements indicate. There are two possible answers to this question, but no clear resolution.

The five priorities listed above are embraced by recommendations COMPLEX has made in the past (Appendix B: [1.13, 1.14, 4.6]). Two of the secondary recommendations put forward in Chapter 12 are consistent with the priorities outlined here.

ASSESSMENT OF PRIORITIES IN THE MARS EXPLORATION PROGRAM

There are no existing plans in the current U.S. Mars Exploration Program to address any of the scientific priorities outlined in the previous section. The Nozomi mission would address items 2, 3, 4, and 5 to some extent. However, that spacecraft has already lost one of its two transmitters, and while one hopes for its success, the mission was not designed for its current 5-year interplanetary cruise phase. At best, it will provide some initial answers to these areas of research, but much more is needed to meaningfully elucidate these open issues. The instruments aboard Mars Express will address issues listed under items 2 and 3 above, but here again, much will be left unanswered. It is also noted that both spacecraft will arrive at Mars during solar minimum, and as indicated above, data from solar maximum are imperative in order to answer some of the outstanding questions (e.g., regarding nonthermal escape).

The instruments needed for a meaningful attack on the five scientific questions listed above would require no new, basic instrument development, and could be installed as a partial payload complement on an orbiting spacecraft. The neutral winds can be measured by either a "baffled" neutral mass spectrometer or a Fabry-Perot interferometer. The latter instrument, along with a good ultraviolet spectrometer, could address in a meaningful way the hot atom and neutral escape flux questions. The neutral mass spectrometer would also provide neutral composition and temperature information. A plasma instrument complement consisting of a magnetometer, a low-energy ion mass spectrometer (capable of measuring flow velocities and temperatures), an electron spectrometer, a plasma wave detector, and a Langmuir probe would go a long way toward resolving issues raised in items 3, 4, and 5 above.

REFERENCES

1. A.O. Nier and M.B. McElroy, "Composition and Structure of Mars' Upper Atmosphere: Results from the Neutral Mass Spectrometers on Viking 1 and 2," *Journal of Geophysical Research* 82: 4341-4349, 1977.
2. C.A. Barth, A.I.F. Stewart, S.W. Bougher, D.M. Hunten, S.J. Bauer, and A.F. Nagy, "Aeronomy of the Current Martian Atmosphere," pp. 1054-1089 in *Mars*, H.H. Kieffer, B.M. Jakosky, C.W. Snyder, and M.S. Matthews (eds.), University of Arizona Press, Tucson, 1992.
3. D.E. Anderson, "Mariner 6, 7 and 9 Ultraviolet Spectrometer Experiment: Analysis of Hydrogen Lyman Alpha Data," *Journal of Geophysical Research* 79: 1513-1518, 1974.
4. C.A. Barth, J.B. Pearce, K.K. Kelly, L. Wallace, and W. G. Fastie, "Ultraviolet Emission Observed Near Venus from Mariner 5," *Science* 158: 1675-1678, 1967.
5. A.F. Nagy, T.E. Cravens, J.-H. Yee, and A.I.F. Stewart, "Hot Oxygen Atoms in the Upper Atmosphere of Venus," *Geophysical Research Letters* 8: 629-632, 1980.
6. L.J. Paxton, "Pioneer Venus Orbiter Ultraviolet Spectrometer Limb Observations: Analysis and Interpretation of the 166- and 156-nm Data," *Journal of Geophysical Research* 90: 5089-5096, 1985.
7. J. Kim, A.F. Nagy, J.L. Fox, and T.E. Cravens, "Solar Cycle Variability of Hot Oxygen Atoms at Mars," *Journal of Geophysical Research* 103: 29339-29342, 1998.
8. A.F. Nagy, M. Liemohn, J.L. Fox, and J. Kim, "Hot Carbon Densities in the Exosphere of Mars," *Journal of Geophysical Research* 106: 21565-21568, 2001.
9. G.M. Keating, S.W. Bougher, R.W. Zurek, R.H. Tolson, G.J. Cancro, S.N. Noll, J.S. Parker, T.J. Schellenberg, R.W. Shane, B.L. Wilkerson, J.R. Murphy, J.L. Hollingsworth, R.M. Haberle, M. Joshi, J.C. Pearl, B.C. Conrath, M.D. Smith, R.T. Clancy, R.C. Blanchard, R.G. Wilmoth, D.F. Rault, T.Z. Martin, D.T. Lyons, P.B. Esposito, M.D. Johnston, C.W. Whetzel, C.G. Justus, and J.M. Babicke, "The Structure of the Upper Atmosphere of Mars," *Science* 279: 1672-1676, 1998.
10. S.W. Bougher, R.G. Roble, E.C. Ridley, and R.E. Dickenson, "The Mars Thermosphere: 2. General Circulation with Coupled Dynamics and Composition," *Journal of Geophysical Research* 95: 14811-4827, 1990.
11. W.B. Hanson, S. Sanatani, and D.R. Zuccaro, "The Martian Ionosphere as Observed by the Viking Retarding Potential Analyzers," *Journal of Geophysical Research* 82: 4351-4363, 1977.
12. W.B. Hanson and G.P. Mantas, "Viking Electron Temperature Measurements: Evidence for a Magnetic Field in the Martian Ionosphere," *Journal of Geophysical Research* 93: 7538-7544, 1988.
13. Y.W. Choi, J. Kim, K.W. Min, A.F. Nagy, and K.I. Oyama, "Effect of the Magnetic Field on the Energetics of Mars' Ionosphere," *Geophysical Research Letters* 25: 2753-2756, 1998.
14. M.H.G. Zhang, J.G. Luhmann, and A.J. Kliore, "A Post Pioneer Venus Reassessment of the Martian Dayside Ionosphere as Observed by Radar Occultation Methods," *Journal of Geophysical Research* 95: 14829-14839, 1990.
15. M.G.H. Zhang, J.G. Luhmann, and A.J. Kliore, "An Observational Study of the Nightside Ionospheres of Mars and Venus with Radio Occultation Methods," *Journal of Geophysical Research* 95: 17095-17102, 1990.
16. R.Z. Sagdeev and A.V. Zakharov, "Brief History of the Phobos Mission," *Nature* 341: 581-584, 1989.
17. W. Riedler, D. Mohlmann, V.N. Oraevsky, K. Schingenschuh, Ye. Yeroshenko, J. Rustenbach, Oe. Aydogar, G. Berghofer, H. Lichtenegger, M. Delva, G. Schelch, K. Pirsch, G. Fremuth, M. Steller, H. Arnold, T. Raditsch, U. Auster, K.H. Fornacon, H.J. Schenk, H. Michaelis, U. Motschmann, T. Roatsch, K. Sauer, R. Schroter, J. Kurths, D. Lenners, J. Linthe, V. Kobzev, V. Styashkin, J. Achache, J. Slavin, J.G. Luhmann, and C.T. Russell, "Magnetic Fields Near Mars: First Results," *Nature* 341: 604-607, 1989.
18. R. Grard, A. Pedersen, S. Klimov, S. Savin, A. Skalsky, J.G. Trotignon, and C. Kennel, "First Measurements of Plasma Waves Near Mars," *Nature* 341: 607-609, 1989.
19. D. Vignes, C. Mazelle, H. Reme, M.H. Acuña, J.E.P. Connerney, R.P. Lin, D.L. Mitchell, P. Cloutier, D.H. Crider, and N.F. Ness, "Solar Wind Interaction with Mars: Locations and Shapes of the Bow Shock and Magnetic-Pile-Up Boundary from the Observations of the MAG/ER Experiment Onboard Mars Global Surveyor," *Geophysical Research Letters* 27: 49-52, 2000.
20. D.L. Mitchell, R.P. Lin, H. Reme, D.H. Crider, P.A. Cloutier, J.E.P. Connerney, M.H. Acuña, and N.F. Ness, "Oxygen Auger Electrons Observed in Mars' Ionosphere," *Geophysical Research Letters* 27: 1871-1874, 2000.

11

Rationale for Sample Return

Chapters 2 through 10 briefly summarize current knowledge of Mars. This chapter has a different purpose: It focuses on the concept of Mars sample return—the immediate goal toward which the Mars Exploration Program is building. Until recently, NASA had been planning for the first element of a sample-return mission to be launched in 2005. Currently, however, the first such mission is to be no earlier than 2011. Although this delay is unfortunate from a scientific perspective, technological and fiscal reasons probably dictate it.

Mars has experienced a complicated history that has created a wide variety of surface and subsurface environments. To select among these during the search for life, much more will have to be known about their origins, histories, and relationships. It will be practically impossible to develop that kind of information at the required level of detail without having samples returned to Earth for study with the full range of laboratory instruments and methodology available here. Many studies have shown the advantages of bringing back samples for study in laboratories on Earth,[1,2,3,4] but it is worth revisiting the issue.

IMPORTANCE OF SAMPLE-RETURN MISSIONS IN THE FRAMEWORK OF NASA'S MARS EXPLORATION PROGRAM

The importance of analyzing returned samples is considered below in two separate sections. The first is concerned with understanding the nature of the samples themselves—for example, their elemental, mineralogical, and isotopic composition. This information addresses science objectives discussed above relating to geochemistry and petrology, chronology, and climate (Chapters 3, 4, and 9, respectively). The second section deals with understanding the nature of any organic or biological material that the samples may contain, topics discussed in Chapter 7. As these preceding chapters show, ongoing developments have consistently raised the priority for early return of samples.

Some martian samples have already been "returned" to Earth. The SNC meteorites (discussed in Chapter 3) have provided both a tantalizing view of a few martian rocks and a demonstration of how much can be learned when samples can be examined in Earth-based laboratories. These meteorites represent, however, a highly selected subset of martian materials, specifically, very coherent rocks of largely igneous origin from a small number of sources. The samples that could provide the most information about martian climate history are something different—namely, sediments and soil samples (SNC meteorites represent the other end of the rock spectrum). Taking Yosemite Valley as a terrestrial analog, the SNC meteorites represent the cliffs rather than the river muds and the sediments from the outwash stretching into California's Central Valley. It is the latter materials that could provide information about timing, chemical conditions, and biological processes, and it is their martian analogs that are sought in sample-return missions.

Geochemistry, Petrology, Chronology, and Climate

Understanding the nature and origin of a rock involves examination of its many properties in great detail, using a variety of techniques. Usually the bulk elemental composition of the rock is determined both for the major elements and for several tens of trace elements that provide strong clues about and constraints on the nature of the differentiation events that led to the formation of the rock. This information is much more valuable when combined with microscopic studies, since rocks contain a nearly infinite amount of information on a microscopic scale, some of it crucial to an understanding of the rock's origin and history. Detailed petrographic examination of the rock is needed to precisely determine the compositions, amounts, and textures of all the minerals present. Measurement of the isotopic composition of a variety of elements in separated mineral grains allows the age of the rock to be determined, and provides constraints on the differentiation history of the system that gave rise to it. Combination and comparison of these chemical, petrographic, and isotopic signals can provide information about relationships between the components of the system.

If the analytical results can be placed in a planetary context, the informational returns can be much greater. For example, it might be shown that the chemical and isotopic composition of a particular mineral indicated clear relationships to a process or source area already recognized elsewhere on the planet. Alternatively, the interpretation of the initial analytical results might have indicated that a fluid—no longer present—had altered the minerals at some point in their history. Either that fluid itself or further evidence for its presence might show up in materials from other sites. As such data accumulate, a detailed understanding of the evolution and significance of a complex rock can be obtained.

Often, in laboratory studies of important extraterrestrial samples, a team of investigators using different analytical techniques is organized into a "consortium" to study a particular rock. Most notable for Mars studies was a consortium organized by J.C. Laul to investigate the Shergotty (Mars) meteorite.[5] The results of this consortium study were published together as a collection of 19 manuscripts in a single issue of a journal, which probably did as much as any other body of research published to date to increase our understanding of the nature of the geochemical processes on Mars. This type of detailed investigation of a sample, where coordinated analyses are made on carefully separated microscopic mineral grains, cannot be done in situ by a robotic spacecraft.

A few related types of measurements can be made in situ, but they do not provide the information needed to thoroughly understand a body as complicated as Mars. For example, it may be possible to determine in situ rock ages based on K-Ar dating, and some researchers believe that these ages may be good enough to calibrate the cratering record on Mars, a very important scientific objective (see Chapter 4 in this report). However, it is known from the study of martian meteorites that their ages are not easily understood. When the data are examined in detail, a complex history of formation followed by multiple disturbances is revealed.[6] This understanding was arrived at only by long and arduous studies of different mineral separates made from rocks that were studied by a wide variety of isotope chronometers. K-Ar dating alone could not have provided it.

Learning about the past climate on Mars is another important objective of Mars science, and returned samples offer the best way to understand an important product of past climates. Ultimately it may be possible to return ice cores from the martian poles that directly address the planet's climate history, but even the first samples collected will contain information about the climate in the layer of weathering products that one expects to find on rock samples. These products will almost certainly be very complex minerals or amorphous reaction products that will tax our best Earth-based laboratory techniques to understand. It is very unlikely that anything but a highly qualitative and ambiguous description of the weathering products could be made by robotic instruments operating on the martian surface.

For these data on weathering products to be most useful for understanding Mars, they need to be considered in the light of remote and in situ observations of stratigraphic layers on Mars (see Chapter 5 in this report). The returned sample will provide a valuable synergism to remote and in situ spacecraft studies, both by providing ground-truth and by allowing for the design of better instruments in the future that are optimized for detection of the properties that samples are found to have.

Biology and Paleobiology

To date, a single set of robotic studies has searched for extant life on Mars: the Viking life-detection experiments, which were designed to test for organisms that used as their carbon source either carbon dioxide or organic molecules. Although the results obtained from the three sets of experiments are regarded as having shown the materials tested to be devoid of both organic compounds and evidence of life,[7,8] this interpretation has been subject to debate.[9] The lack of agreement highlights the difficulties inherent in the detection of viable microorganisms by robotic means. Indeed, even were there unanimity that the Viking experiments did not show the presence of life, the experiments could still be criticized as being overly "geocentric" in that they showed a lack of evidence of metabolism only of those types particularly common among terrestrial microbes, not of all conceivable metabolisms (nor even of various redox-reaction-based microbial metabolisms well known on Earth). Moreover, the problem of distinguishing between biological and nonbiological organic material is complicated. The carbonaceous chondrites, interplanetary dust particles, and probably other bodies within the solar system contain abundant organic material that is structurally similar to biological products. Definitive resolution of the differences between biotic and abiotic organic molecules requires highly sophisticated techniques well beyond any that could be managed robotically.

Thus, at the present state of knowledge, results obtained from any life-detection experiment carried out by robotic means seem likely to be ambiguous: (1) results interpreted as showing an absence of life will be regarded as too geocentric or otherwise inappropriately limited; (2) results consistent with, but not definitive of, the existence of life (e.g., the detection of organic compounds of unknown, either biological or nonbiological, origin) will be regarded as incapable of providing a clear-cut answer; and (3) results interpreted as showing the existence of life will be regarded as necessarily suspect, since they might reflect the presence of earthly contaminants rather than of an indigenous martian biota. Finally, the detection of life robotically is unlikely to be accomplished by a search for either of the two categories of fossil life that might be brought to bear on the problem: stromatolites and microfossils.

Formally defined, a stromatolite is an accretionary organosedimentary structure, commonly thinly layered, produced by the activities of mat-building communities of mucilage-secreting microorganisms (see Figure 11.1). Unfortunately, however, on Earth true stromatolites can be confused with nonbiologically deposited look-alikes—cave rocks, such as stalactites and stalagmites, and hot spring deposits such as those formed where minerals build up in thin, sometimes wavy layers as they crystallize from solution. On Earth, microbes are so widespread that there is practically no place where stromatolitic look-alikes form without life playing at least a minor role. But on a planet where life never got started, there could be many places veneered by thinly layered stromatolite-like deposits unrelated to life—laid down, for example, by repeated wetting and drying or freezing and thawing of mineral-charged salt pans or shallow lagoons. Moreover, it is useful to recall that stromatolites were known on Earth for more than a century before their microbial origin was firmly established.[10] Were stromatolite-like structures to be photographed on the surface of Mars, it seems certain that there would be widespread questioning as to whether the objects detected were in fact produced by life.

In a similar vein, it seems unlikely that robotic detection of fossil-like objects in, or on the surfaces of, rocks on Mars would prove sufficiently convincing to demonstrate to an acceptable level of certainty that past life existed on that planet. Although it is likely that optical studies of robotically prepared petrographic thin sections could overcome problems of establishing whether objects detected are indigenous, the crucial problem of demonstrating the biogenicity of the objects in question would remain. Here lessons learned from the search for ancient (Precambrian) microbes on Earth would certainly apply—lessons that well illustrate the error of assuming that microstructures "unlike known mineral forms" can be regarded as "fossils" simply for want of any other explanation, as they have been repeatedly in the past.[11,12]

In summary, at the present state of knowledge and technological expertise—and, probably, for the forthcoming several decades—it is unlikely that robotic in situ exploration will prove capable of demonstrating to an

FIGURE 11.1 Fossilized stromatolites (age, 500 million years) in Saratoga Springs, New York. Image available online at <http://www.petrifiedseagardens.org>. Photograph courtesy of Joseph Deuel, Petrified Sea Garden, Inc.

acceptable level of certainty whether there once was or is now life on Mars. For the foreseeable future, it is reasonable to expect that such studies can be performed only on samples retrieved from Mars and brought to Earth for detailed investigation in appropriately equipped laboratories.

GENERAL CONSIDERATIONS

As noted, samples from Mars are already on Earth in the form of SNC meteorites. These are, however, a very highly selected and probably altered subset of all the possible Mars samples. They include only massive, coherent lithologies that carry no information about martian surface processes. The SNC meteorites were launched from Mars by impact processes with resultant shock and heating, then drifted in space (where possibly they were involved in numerous secondary collisions), and finally fell through Earth's atmosphere. As samples of Mars they are better than nothing, but they are far from optimal research material.

There are very important advantages to collecting samples on Mars and bringing them to Earth for study. One stems from the ability to look at ever-smaller pieces of samples, as instrumental sensitivities continue to improve. For example, it is now possible, using an ion microprobe, to determine the isotopic composition of a mineral zone or fragment as small as a micron across. It is known from studies of the Apollo lunar samples that often one can find samples of rocks present as soil particles or in breccias (compacted soils) that are derived from inaccessible parts of the planet. A single regolith sample can contain tiny samples of rocks from widely different locations on the planet's surface. As an example, Wood and colleagues were able to infer the composition of the lunar highlands from rock fragments found in an Apollo 11 soil sample before a highlands site had actually been visited.[13] Researchers' capacity to analyze very small subsamples also means that a returned Mars sample of modest size can be divided and studied by a large number of scientists in laboratories with diverse capabilities.

Additionally, any investigation into an alien world tends to uncover as many questions as it answers. Having samples present on Earth allows investigators to develop and to answer refined, second-order questions. It also makes it possible for observations of particular importance to be checked by multiple investigators using the same or different techniques.

Though robotic missions that perform in situ analyses will continue to add incrementally to the knowledge of Mars, their advances relative to those provided by returned samples will be minor. Even if the first returned samples are not optimal in terms of siting, they will provide a greatly enhanced view of the geologic processes on Mars. Even a grab-sample of soil from a randomly chosen site on the planet will reveal the character of martian surface material: its chemistry, oxidation state, content of organic materials, mineralogy, and the history of weathering reactions that has affected it. Also, the properties of the ubiquitous martian dust will be determined, information that will allow corrections to be applied to the data sets of past and future robotic orbital and lander missions. Detailed knowledge of the surface material will permit a more intelligent choice of measurements to be made by future robotic missions. (None of this information is available from SNC meteorites, which are not surface samples.) Rocks collected at a randomly chosen site will be suitable for study by a variety of isotopic and chemical techniques that will reveal the nature and chronology of the planetary fractionating events that produced them. They will also contain an isotopic record of the integrated effects of the martian surface radiation environment.

The exercise of even a modest amount of selectivity in landing sites will open additional doors: samples from a formerly fluvial environment, for example, may be found to include rocks with diverse compositions and ages, sedimentary rocks that contain a record of aqueous activity on the martian surface, conceivably even fossil evidence of life.

Observations made by robotic orbiters and landers can provide no more than tantalizing hints and glimpses of the information we want: answers to the questions of whether life ever started on Mars, what the climate history of the planet was, and why Mars evolved so differently from the way Earth did. The definitive answers to these questions will come from the study of Mars samples in laboratories on Earth.

REFERENCES

1. M.J. Drake, W.V. Boynton, and D.P. Blanchard, "The Case for Planetary Sample Return Missions: 1. Origin of the Solar System," *Eos* 68(8): 105, 111–113, 1987.
2. J.L. Gooding, M.H. Carr, and C.P. McKay, "The Case for Planetary Sample Return Missions: 2. History of Mars," *Eos* 70: 745, 754–755, 1989.
3. G. Ryder, P.D. Spudis, and G.J. Taylor, "The Case for Planetary Sample Return Missions: 3. Origin and Evolution of the Moon and its Environment," *Eos* 70: 1495, 1505–1509, 1989.
4. T.D. Swindle, J.S. Lewis, and L.A. McFadden, "The Case for Planetary Sample Return Missions: 4. Near-Earth Asteroids and the History of Planetary Formation," *Eos* 72: 473, 479–480, 1991.
5. J.C. Laul, "The Shergotty Consortium and SNC Meteorites—An Overview," *Geochimica et Cosmochimica Acta* 50: 875–887, 1986.
6. H.Y. McSween, "The Rocks of Mars, From Far and Near," *Meteoritics and Planetary Science* 37: 7–25, 2002.
7. K. Biemann, J. Oró, P. Toulmin III, L.E. Orgel, A.O. Nier, D.M. Anderson, P.G. Simmonds, D. Flory, A.V. Diaz, D.R. Rushneck, J.E. Biller, and A.L. Lafleur, "The Search for Organic Substances and Inorganic Volatile Compounds in the Surface of Mars," *Journal of Geophysical Research* 82: 4641–4658, 1977.
8. H.P. Klein, "The Viking Mission and the Search for Life on Mars," *Reviews of Geophysics and Space Physics* 17: 1655–1662, 1979.
9. G.V. Levin, and P.A. Straat, "Viking Labeled Release Biology Experiment: Interim Results," *Science* 194: 1322–1329, 1976.
10. J.W. Schopf, *Cradle of Life: The Discovery of Earth's Earliest Fossils*, Princeton University Press, Princeton, N.J., 1999.
11. J.W. Schopf and M.R. Walter, "Archean Microfossils: New Evidence of Ancient Microbes," pp. 214–239 in *Earth's Earliest Biosphere, Its Origin and Evolution*, J.W. Schopf (ed.), Princeton University Press, Princeton, N.J, 1983.
12. C.V. Mendelson and J.W. Schopf, "Proterozoic and Selected Early Cambrian Microfossils and Microfossil-like Objects," pp. 865–951 in *The Proterozoic Biosphere, A Multidisciplinary Study*, J.W. Schopf and C. Klein (eds.), Cambridge University Press, New York, 1992.
13. J.A. Wood, U.B. Marvin, B.N. Powell, and J.S. Dickey, Jr., "Lunar Anorthosites," *Science* 167: 602–604, 1970.

12

Assessment of the Mars Exploration Program

INTRODUCTION

Chapters 2 through 10 above review the state of knowledge of the planet Mars. Recent missions, and in particular the Mars Global Surveyor (MGS) mission, have made surprising and important new discoveries and have greatly increased our understanding of the planet. Several scientific issues have been resolved as a result of the new information, and new directions of inquiry have opened up. This chapter summarizes the science highlights and shows the extent to which the Mars science priorities will be met by currently planned U.S. and foreign missions. Following the discussion of the importance of sample return in Chapter 11, this chapter also discusses aspects of the sample-return missions toward which the U.S. program is building, as well as several other issues affecting the Mars Exploration Program.

The evolution of thought about Mars science priorities since 1978 has been methodical and thorough. The science priorities recommended by COMPLEX and other groups remain fully valid in light of the new discoveries, although the degree to which NASA and other space agencies have responded to the priorities has been uneven. The new NASA Mars Exploration Program (MEP), announced in October 2000, is a science-driven program that seeks to understand Mars as a dynamic system and to understand whether life was ever part of that system. As part of this program of global understanding, NASA's goals emphasize:

- Understanding the martian climate;
- Understanding the role water plays in the environmental history of Mars;
- Understanding Mars's biological potential and its connection to the climate record; and
- Understanding the interactions between the surface, atmosphere, and interior and how they are related to water.

The NASA philosophy "Seek, In Situ, Sample" is designed to advance learning about the global dynamics of the Mars system and to narrow down the search to focus on its biological potential. The MGS and Mars Odyssey missions were designed for global reconnaissance, and the Mars Exploration Rovers (2003) and Mars Reconnaissance Orbiter (2005) missions, which will carry out field geology and remote sensing, will focus on finding environmental indicators that suggest the possibility of life. These missions will identify a suite of sites for intensive surface analysis by Mars Science Laboratory[a] (2007),[b] in principle permitting the best sites for sample return to be located.

Overall the "Seek, In Situ, Sample" strategy is a sound one, and NASA has built up a strong, risk-attentive program focusing on the understanding of Mars. This is an excellent approach, and in the areas that focus on answering questions about past or present life, NASA is doing a good job of addressing priorities for Mars science that have been recommended by COMPLEX and other groups since 1978. However, it is important to insert a strong note of caution regarding this plan. The effort must focus on answering the question, Did life ever arise on Mars?, and not on searching for evidence of the life that "must surely be there." If life ever arose, its signatures may only be evidenced through detailed geochemical or isotopic investigations, and separation of biotic from abiotic signatures will require an extensive understanding of all of Mars's systems and their histories. From that perspective, COMPLEX notes that the current NASA strategic plan does not address all of the priorities needed to understand Mars as a dynamic system suitable for life.

NEAR-TERM MISSIONS—MARS RESEARCH OPPORTUNITIES

NASA's near-term Mars missions (presented in Table A.1 in Appendix A) includes a lineup of vehicles and investigations that will address a wide range of important scientific goals, while remaining within the program's overall budgetary constraints. A summary of the priority science that will be met by these missions appears below. Non-U.S. missions are also included in the list, as they are complementary to NASA's missions. A later section of this chapter contains COMPLEX's assessment of MEP and its congruence with recommended science priorities.

1998—Nozomi (Japan)

The Japanese Nozomi mission (formerly called Planet B) carries 14 scientific instruments (including a NASA neutral mass spectrometer) designed to investigate the structure of Mars's upper atmosphere, ionosphere, magnetic field environment, and solar wind processes. Because the spacecraft used up too much fuel in a trajectory correction maneuver, it will arrive at Mars in 2004, 5 years later than originally planned (see Chapter 10 in this report). The mission was not designed to be long-lived, and some of the spacecraft systems and instruments may suffer damage during the long and unplanned cruise. Nozomi is one of the few missions to address important upper-atmospheric science priorities; unfortunately, it is considered to be compromised. None of the near-term NASA missions has upper atmospheric/ionospheric mission objectives.

2001—Mars Odyssey

Mars Odyssey, an orbiter, carries a payload consisting of the GRS instrument, which was part of the original Mars Observer payload, THEMIS, and MARIE (a radiation environment experiment that will characterize Mars's galactic cosmic radiation environment). GRS should provide valuable global-scale maps of Mars's elemental composition with a spatial resolution of 300 km. It will also produce maps of the near-surface hydrogen abundance, which should be interpretable in terms of the abundance of water ice and adsorbed water in the near-surface regolith. The THEMIS instrument is a high-spatial-resolution multichannel infrared radiometer and imager. The primary goal of THEMIS is to map the global surface emissivity of Mars in 10 spectral bands at a spatial resolution of 100 m, much higher than has been accomplished by previous infrared instruments such as Mariner 9's Infrared Interferometer Spectrometer and MGS's TES. THEMIS may be able to identify some minerals as well, but for the most part it will define broad surface emissivity and petrologic units that will be useful for choosing future landing sites. The imager on THEMIS will acquire 20-m-spatial-resolution images in five wavelength bands covering the visible to near-infrared region, but of only 5 to 10 percent of the planet. The GRS will provide a critical global data

[a]Also referred to as the Mars Smart Lander, the Mobile Science Laboratory, and by a variety of other names.

[b]Following the completion of this study, NASA announced that it was delaying the launch of the Mars Science Laboratory until 2009 to allow time to develop an advanced, radioisotope power system for this mission.

set, and THEMIS will be important for scaling thermal data from the surface to orbit and for identifying mineralogically interesting landing sites. However, considering the difficulty experienced in extracting mineralogical information from the data obtained by previous and ongoing orbital infrared experiments, the results from THEMIS may not be as definitive as expected. The mission will also help refine the record of geologic stratigraphy on smaller scales than was possible before.

2003—Mars Exploration Rovers

The two Mars Exploration Rovers (MERs) scheduled for launch in June 2003 will each carry the Athena integrated payload, which had originally been selected for the 2003 sample-collection mission. The MERs will provide improved surface mobility relative to Mars Pathfinder's Sojourner rover, as well as a significantly enhanced capability for characterizing rock mineralogy. A goal is to link chemistry and mineralogy at the surface with that surmised from orbital observations and to find conclusive evidence of water-affected surface materials, and thus regions where conditions may have been favorable to the preservation of biotic processes. One concern about MER is that its landing ellipse is too large and its roving range is too small to guarantee access to the most exciting geological sites that have been identified in MGS data. This, combined with engineering concerns regarding landing safety, may ultimately result in the selection of landing sites that limit the overall scientific potential of the 2003 missions. (Some landing ellipses under consideration have interest that is regional in extent; in these, limited rover range is less damaging.)

2003—Mars Express (European Space Agency)

The payload of the European Space Agency's Mars Express represents an attempt to recover some of the science lost with the failure of Russia's Mars 1996 mission. The payload includes the HRSC, which will obtain global color images at resolutions between 10 and 30 m and topography at the same spatial resolution, and 2-m images of 1 percent of the planet. Global observations of visible/near-infrared reflected light at an average resolution of 1 km will be acquired by OMEGA and will be used to investigate surface mineralogy and the atmosphere. Global circulation measurements; high-resolution mapping of atmospheric composition (including water vapor); density, temperature, and pressure profiles in the atmosphere; and interaction of the atmosphere with the interplanetary medium will be studied by the Planetary Fourier Spectrometer (PFS), SPICAM, ASPERA, and the Radio Science Experiment (MaRS). In addition, MARSIS will map subsurface structures at kilometer scale, looking for subsurface liquid and solid water.

2003—Beagle 2 (United Kingdom)

The United Kingdom's Beagle 2 is a small but heavily instrumented spacecraft, and it represents the landed component of Mars Express. The mission goals are to examine the geology, surface composition, oxidation state, and mineralogy of rocks at the landing site and to make detailed atmospheric-composition measurements, including determination of isotopic fractionation. These exobiology measurements will be made with a suite of instruments including panoramic, wide-angle, and microscope cameras, a gas chromatograph-mass spectrometer, a Mössbauer spectrometer, an alpha-proton-x-ray spectrometer, and an environmental surface weather station.

2005—Mars Reconnaissance Orbiter

Currently in the early stages of development, Mars Reconnaissance Orbiter (MRO) promises to fill a number of gaps in our knowledge concerning the search for compelling environmental indicators related to the action of liquid water and possible biological processes as well as to provide significant new discoveries. Instruments that are expected to be on the payload are the PMIRR and MARCI instruments from the failed Mars Climate Orbiter (MCO), an extremely high resolution imager (called HiRISE) with a surface resolution of tens of centimeters, a visible/near-infrared mapping spectrometer (known as CRISM) with a spatial resolution of 50 m, and SHARAD, a radar sounder to be provided by the Italian space agency which will acquire multiple vertical atmospheric profiles over 1 martian year.

The purpose of this mission is to recover the critical atmospheric measurements lost with Mars Observer and Mars Climate Orbiter, and to investigate in detail hundreds of potential landing sites with high biological potential, using the high-resolution, high-data-rate instruments. With its polar orbit and high-resolution imaging capability, this is one of the few NASA missions with the ability to obtain information on climate, by looking at the polar layered terrain.

2007—Mars Science Laboratory[c]

The advanced rover mission known as the Mars Science Laboratory (MSL) is now in the definition stage. The goal of this mission is to develop technologies required for future sample-return missions, as well as to conduct landed science focusing on surface and subsurface materials potentially linked to life. The MSL will be a pathfinding mission to the most biologically interesting sites. The lander's engineering goals of improved landing precision, terminal hazard avoidance, high rover mobility, and long surface lifetime are responsive to the needs of sample-return missions. The main concerns relate to whether all technical challenges can be met in addition to the challenges of the science activities, and to the limitations imposed by the solar power supply. (See the subsection "Power Supply for Landers and Rovers" below in this chapter.) One experiment that deserves cautious consideration is a proposed drilling system designed to provide subsurface access to depths of 2 m. Such a system could consume a large fraction of the mission's resources, and it may not be compatible with the goal of high mobility.

2007—Mars Scout

A principal-investigator-led Mars Scout mission will be selected in 2003 for a launch opportunity as early as 2007. The Mars Scout program offers NASA an excellent opportunity to fill gaps in our knowledge, to react to recent scientific discoveries, and to increase community involvement. As such, Mars Scout needs to be open to all science investigations of Mars that are not addressed within the scope of the existing program. Topics beyond the current emphasis on "water and life" should be encouraged. (See the subsection "The Scout Program," below in this chapter.)

2007—NetLander (France)

The Mars NetLander mission is sponsored by a consortium of European nations led by France. The mission objective is to perform simultaneous surface measurements in order to investigate Mars's internal structure, its subsurface layering, and the atmosphere. During the baseline 1 martian year mission, the NetLander payloads will conduct simultaneous seismic, atmospheric, magnetic, and ionospheric measurements, with a surface meteorology package, a ground-penetrating radar, an electric field package, a magnetometer, the NetLander Ionosphere and Geodesy Experiment, a panoramic camera, a seismometer, and a soil properties package (SPICE). This is the one near-term mission that significantly addresses issues related to the deep interior of Mars. Another important mission objective is to search for subsurface reservoirs of liquid or frozen water.

The Longer Term—Sample-Return Missions

Returned samples from Mars will be extremely powerful tools for answering many of the most important scientific questions about the planet and its history (see Chapter 11). In 1995—prior to the ALH84001 Mars meteorite discoveries in 1996—NASA had begun planning a mission to collect and cache Mars samples, to be

---

[c]Following the completion of this study, NASA announced that it was delaying the launch of the Mars Science Laboratory until 2009 to allow time to develop an advanced, radioisotope power system for this mission.

launched in 2003, even before a plan to convey the samples back to Earth was fully formulated. As events have unfolded, the prospect of sample return has receded farther and farther into the future. According to the current program (2002), we are more than a decade away from the first potential launch of a sample-return mission. This scaling back of ambitions aptly demonstrates that Mars sample-return missions are nontrivial in their engineering scope, cost, and risk, and that they will require significant investments in technology, and possibly international cooperation, to be successful.

The results of the analysis of Mars Pathfinder and MGS observations provide some potential guidance in sample return strategies. Before those missions, it was considered sufficient to target landers to "grab bag" landing sites that were hoped to contain a wide variety of rocks that could be accessed at a single location, with small requirements for mobility. Now it appears that Mars's geologic and climatic history is best exposed in widely separated, isolated locations, and that a complete picture of the planet's history probably cannot be obtained from samples collected at a single location. Instead, a series of samples will be required from diverse locations on the planet, and this will require multiple sample-return missions to accomplish (discussed below in the section "Mars Sample Returns"). Thus, the first sample return should be seen as a "pathfinder" for future sample-return missions, and it should be used to develop the key technologies, procedures, and infrastructure necessary to embark on a future program in which samples are returned from many locations on the planet.

STATUS OF HIGH-PRIORITY SCIENCE QUESTIONS

Mars Pathfinder and Mars Global Surveyor have greatly improved our knowledge of the interior of Mars (see Chapter 2 in this report). The former gave us a significantly improved moment-of-inertia factor, leading to a better estimate of the bulk composition of the planet. Topographic data from MGS have shown that the Tharsis Plateau region predates the fluvial channels, opening up the possibility that volcanic gas release from the plateau's formation would have created suitable conditions for the creation of the channels. The topographic data likewise ruled out the impact hypothesis for the hemispheric dichotomy. In order to fully understand the interior structure and composition, however, knowledge of the moment-of-inertia factor must be coupled with the core size, which is still a free parameter. When it is known, the bulk composition of Mars will be much more closely constrained, which will have important implications for the raw material of Mars and cosmochemical models for the pressures and temperatures in the solar nebula that gave rise to that raw material. We will understand whether Mars has an oxidation state higher than that of Earth, and if the temperature and pressure in the accretion zone for Mars allowed significantly different mineral condensation there.

The composition of Mars's surface materials is coupled with the interior structure of the planet, yet recent missions have provided information only about the major element compositions of rocks and fines at a few sites (see Chapter 3 in this report). Beyond the suggestion by MGS's TES that a distinction can be drawn between basaltic and andesitic areas of the Mars surface, we are largely ignorant of the surface mineralogy. TES data provide a tantalizing glimpse of an area of gray hematite near the equator suggestive of large-scale water interactions, but no direct measurements of hydrated minerals exist to date because of insufficient resolution.

The Mariner through MGS missions have shown that water has played a significant role in the evolution of the planet, from evidence of standing water, to large outflow channels, to valley networks, and most recently very youthful channels (see Chapter 6). The higher resolution of the MGS images has returned contradictory evidence of a northern hemisphere ocean and has made the understanding of the valley networks more complex, since the data suggest that the sources of the eroding fluid are not from surface runoff. The discovery of channels on very steep slopes makes the hypothesis of possible subsurface water sources more complex, and has suggested the possibility that channels were created by other volatiles.

The MGS discovery of remnant crustal magnetic anomalies suggests an early dynamo, but because of the lack of absolute dating of the surface, it cannot be determined when the dynamo ceased (see Chapter 2). Similarly, while the MGS mission established a good relative stratigraphic record for Mars, models using the lunar cratering curve can link this to absolute ages only to within a factor of two (see Chapter 4). This is insufficient accuracy for understanding the global dynamic system of interactions among the interior, surface, hydrosphere, and atmosphere during the periods of most interest to the question of the search for life.

Many of the advances in the understanding of the potential for life on Mars have come not from recent missions but rather from studies of life in extreme environments on Earth and from studies of the SNC meteorites. On the basis of those studies, coupled with the evidence from MGS and earlier missions that the past climate on Mars was very different from what it is now, a better understanding is being built of the types of environments on Mars that could constitute suitable life habitats. MGS data have shown striking detail in the polar deposits (see Chapter 9), hinting that there may be records of quasi-periodic climate variations recorded there. Surface features give clear evidence of past climate differences, but it is unknown if there was a sustained warm, wet climate or just episodic nonequilibrium events. These questions are crucial to future progress in understanding Mars as a habitable planet.

A key to understanding the past climates is an understanding of the lower and upper atmospheres and their evolution (see Chapters 8 and 10). The Viking and MGS missions have provided a good estimate of global seasonal water vapor variations but have been poor at monitoring daily variations. Similarly, MGS, Mars Pathfinder, Viking, and Hubble Space Telescope have provided much data on dust loading in the atmosphere. This information has given us a good basic understanding of the pole-to-pole general circulation patterns, but the seasonal circulation patterns are unknown, and the evolution of the atmosphere will not be understood until we know the crustal history and composition, the interaction of volatiles with the near-surface layer, and the loss mechanisms at the top of the atmosphere.

Table 12.1 summarizes the outstanding science issues connected with Mars exploration that are discussed in Chapters 2 through 10.

MARS SCIENCE PRIORITIES AFTER MARS GLOBAL SURVEYOR

The Mars science priorities recommended by seven NRC reports since 1978 and by two NASA reports—the 1996 Mars Expeditions Strategy Group report and the 2000 MEPAG report—are assembled in Appendix B. The recommendations reprinted in Appendix B are discussed in Chapters 2 through 10 and are summarized in Table 12.2. Organized in the same sequence and under the same topics as discussed in Chapters 2 through 10, the science priorities are grouped by subject in Table 12.2, proceeding from the interior of the planet outward. Solid circles in the column titled "Panel Recommending" identify the questions that are recommended for study in each report. The column in Table 12.2 labeled "Inclusion in Missions" shows which missions will address these questions. (Solid circles signify missions that will concentrate on each science objective, and open circles signify a lesser level of attention to that objective.) Missions in NASA's Mars Exploration Program are listed separately from the missions projected by other nations.

ASSESSMENT OF NASA'S MARS EXPLORATION PROGRAM AND ITS CONGRUENCE WITH RECOMMENDED SCIENTIFIC PRIORITIES

Table 12.2 shows a high degree of complementarity between the NASA missions and those of other nations. Table 12.2 shows that in many but not all cases, planned missions (foreign as well as NASA) will address the high-priority Mars science issues. However, understandably because of budget constraints, some important areas of Mars science will remain underserved:

- The Mars Exploration Program recognizes the importance of gaining information about the surface, atmosphere, hydrosphere, and interior of the planet to arrive at an understanding of the Mars dynamic system and a global context for assessment of the biological potential of Mars. However, NASA has no plans for missions that address high-priority questions about the interior of Mars.
- Absolute dating of planetary surfaces by isotopic techniques is not contemplated for any NASA or foreign mission prior to sample return, yet understanding the absolute chronology attached to the Mars stratigraphic record is essential to understanding the history and evolution of the planet, water, and climate, and the geological context of the sites visited.
- There is an absence of NASA missions that specifically address Mars's atmosphere, climate, polar science, ionosphere, and solar wind interactions. Direct measurements related to volatiles—for example, the distribution

and behavior of near-surface water—are needed. Some but not all of these goals will be addressed by upcoming foreign Mars missions such as Nozomi, Mars Express, Beagle 2, and NetLander. COMPLEX urges NASA to continue its support for U.S. participation in Mars missions conducted by NASA's international partners.

COMPLEX has identified 10 measurements that could be made, which would contribute knowledge in the underserved areas outlined. In most cases it is fairly clear why these measurements have not been adopted by the Mars Exploration Program. The first four are considered most important:

- *A gas mass spectrometer in a long-lived surface package, to study the chemical dynamics and isotope ratios of C, H, O, and noble gases in the Mars atmosphere (see Chapter 8 in this report).* This would greatly refine the measurements of the Viking aeroshell mass spectrometer. Time variability of isotopic compositions could be interpreted in terms of sources, sinks, and reservoirs of volatiles, and atmospheric evolution.

  Contra this concept: A related experiment is included on the 2003 Beagle 2 mission. There is also a potential problem in providing power to a surface package for a long time; see the subsection "Power Supply for Landers and Rovers," below.
- *Passive seismometers to study the interior of the planet (see Chapter 2).* This has been consistently endorsed by advisory groups since 1978, but it is not addressed in the current NASA Mars Exploration Program.

  Contra: A passive seismology experiment is projected for the 2007 NetLander mission, described above and in Chapter 2; again, there is a potential problem in providing power to seismic stations long enough to register naturally generated seismic signals—see "Power Supply for Landers and Rovers," below.
- *In situ age determination of rocks at selected sites (see Chapter 4).* Even if only a low order of accuracy of radiometric dates (±20 percent, by the ^{40}K-^{40}Ar technique) is achievable, more rocks can be (approximately) dated in situ than through sample return, increasing the potential for calibrating the cratering rate and the stratigraphic column on Mars. An understanding of the evolution of the planet's surface—the timing of events and the coupling of different systems—requires that an absolute chronology be established with significantly better than the factor-of-two uncertainty that now exists for surfaces of intermediate age.

  Contra: More MER/MSL-type missions than are currently projected would be needed to fully exploit this technique; further, it has not been proven that robotic age-dating can be made to work at all.
- *Global mapping of subsurface ice and water in the upper crust by radar (see Chapter 6).* The distribution of ice and water in the crust is consistently identified as one of the most important goals of Mars studies. In principle, it can be learned by pulsing the surface with radar at different frequencies from orbit, to detect materials with contrasting dielectric properties.

  Contra: The cost would be high and the results might be equivocal, i.e., not subject to unique interpretation. Mars Express's MARSIS and Mars Reconnaissance Orbiter's SHARAD experiments will constitute a feasibility study of this concept.

Six other experiments that COMPLEX deems worthy of consideration are these:

- *Long-lived landed humidity sensors.* These would study the exchange of atmospheric water with the regolith.
- *Further mapping of the magnetic field of Mars, to fill in gaps in the MGS magnetic map, along with plasma mapping.* However, these require an elliptical orbit, which is not ideal for other orbital science.
- *Combined calorimetry and precision gas analysis of surface material in the polar regions.* This may be the only opportunity to learn in detail about polar materials, which probably will not soon be targeted for sample return.
- *Measurements from orbit of the dynamics of the middle and upper atmosphere of Mars, and atmospheric escape, using a Fabry-Perot interferometer.*
- *Detection of ice in the martian regolith by neutron spectroscopy (epithermal neutrons from H), using an airborne instrument.*
- *Synthetic aperture radar investigations of the surface to quantify roughness and particle size, and to penetrate below the surficial dust layer to reveal buried morphologies.*

TABLE 12.1 Outstanding Mars Exploration Science Issues

Topic	Priority Issues	Current Plan	Future Possibilities
1. Deep interior	Size of the core Interior activity Solidity of core	Passive seismic network deployed by NetLander	
Crust	Thickness Structure	None	Active seismic experiments
Heat flow	Geothermal gradient	None	Thermal probes in drilled sample holes
Gravity field	More detailed gravity map than that of MGS	Measured by MRO	
Magnetic field	Complete survey of spatial distribution	None	Low altitude magnetic mapping
	Rock magnetization	None	In situ rover magnetic field and mineralogy Study of returned samples
2. Geochemistry, petrology	Rock compositions at selected localities	MER, Beagle 2, MSL	Extend coverage of sampling Study of returned samples
	Rock compositions beneath near-surface altered zone	None	Drilled samples
Areal geochemistry	Outline geochemical provinces, relate to rock compositions	Measurements by MO, ME, MRO	
3. Chronology, stratigraphy	Crystalline rock ages	None	Dating of returned samples, rocks in situ
	Tie cratering record to dated surfaces	Crater record improved by MGS, ME, MRO, CNES Orbiter, but dependent upon rock dating	Dating of returned samples, rocks in situ
	Tie stratigraphic column to dated samples	Stratigraphic data from MGS, ME, MRO, CNES Orbiter, but dependent upon rock dating	Dating of returned samples, rocks in situ
4. Surface processes	Effects of water	Detailed MGS, ME, MRO observations of channels Correlation with rock compositions by MER, Beagle 2, MSL	Study of returned samples
	Effects of wind	Detailed observations of eolian deposits by MGS, ME, MRO	Study of returned samples
	Volcanism	Detailed MGS, ME, MRO observations of volcanic morphology Correlation with rock compositions found by MER, Beagle 2, MSL	Study of returned samples
	Impact cratering	MGS, ME, MRO cratering record improvement	
	Surface alteration	MER, Beagle 2, MSL in situ observations Correlation with spectral mapping by MO, ME, MRO	Study of returned samples

TABLE 12.1 Continued

Topic	Priority Issues	Current Plan	Future Possibilities
5. Water	Water in the atmosphere	Measurements by MO, ME, MRO, NetLander	Reflight of MVACS payload from lost Mars Polar Lander
	Near-surface water in the planet	GRS on MO	Reflight of MVACS payload from lost Mars Polar Lander
	Deep water in the planet	Radar Sounder on ME, MRO, NetLander	Active seismic and electromagnetic studies
	Evidence of water in the past	Detailed observations of channels by MGS, ME, MRO	Study of returned samples
6. Life	Extant life	None	Study of returned samples
	Fossil life	None	Sample return (especially sedimentary rocks)
	Organic material, oxidants	MER, Beagle 2, MSL	Study of returned samples
7. Atmosphere	Composition	ME (PFS, SPICAM) Beagle 2 (ATMIS, GAP) MRO (PMIRR-MkII, MARCI)	In situ isotopic composition surface material measurements Diurnal compositional measurements from orbit Meteorology network; Mars surface upward-looking spectroscopic measurements
	General circulation	Earth-based observations Meteorology network	
	History	From geologic and geochemical evidence (see 3, 4, 6, and 8)	
8. Climate change	Interannual variability	ATMIS on NetLander	Long-lived surface stations
	Quasi-periodic variability	High-resolution MRO observations of layered polar deposits	Landed studies of polar layered deposits
	Long-term climate change	Stratigraphic studies (see 3 above)	Geomorphic studies (see 4 above) Mineralogic studies (see 2 above) Study of returned samples
9. Thermosphere, exosphere, ionosphere	Dynamics of upper atmosphere	None	Optical remote sensing or in situ measurements of winds from a future (to be determined) orbiting spacecraft
	Escape of hot atoms	Nozomi, ME	Nozomi, a compromised mission; additional measurements needed
	Escape of ions	Nozomi	Nozomi compromised; need additional measurements
	Interaction of crustal and interplanetary magnetic field	Nozomi	Nozomi compromised; need additional measurements
	Energetics of the ionosphere	Nozomi	Nozomi compromised; need additional measurements

TABLE 12.2 Comparison of Recommended Science Priorities with Experiments on Projected Flight Missions

Science Priorities	Panels Recommending											Inclusion in Missions: NASA						Inclusion in Missions: Other			
	COMPLEX 1978	COMPLEX 1990	COMPLEX 1990	COMPLEX 1994	NASA 1995	COMPLEX 1996	McCleese 1996	COMPLEX 1996	COMPLEX 1998	NASA 2000	MEPAG 2000	MGS 1997	MO 2001	MER 2003	MRO 2005	MSL 2007	Sample Return	Nozomi 1999	Mars Express 2003	Beagle 2 2003	NetLanders 2007
Interior																					
• What is the size and state of the core?	●		●	●							●										●
• Is Mars active (interior activity, tectonics, volcanism)?				●							●										●
• What is the thickness/structure of the crust?	●		●							●	●	○									
• What is the geothermal gradient?	●										●										
• What is the character/origin/evolution of the magnetic field?	●		●								●	●									○
Geochemistry and Petrology																					
• What variations of geochemistry and petrology are present?	●												●			○	○		●		
• What have been mechanisms of geochemical differentiation?		●															○				
• Is there evidence for aqueous mineralization?					●								○	○	○	○	●		○	○	○
Chronology and Stratigraphy																					
• What are the relative ages of geological units and events?	●	●									●	○	○		○						
• What are the absolute ages of geological units and events?	●	●									●									○	
• What are the absolute ages of crystalline rocks?	●										●						●			○	
Surface Processes																					
• What are the present rates of erosion and deposition?											●	○	○		○				○		
• What were the past rates and processes: water and eolian?	●			●								○	○		○				○		
• What has the role of impact cratering been?												○	○		○				○		
• What role has volcanism played in surface evolution?											●	○			○				○		
• Surface/atmosphere interaction: what volatile sources/sinks?	●										●	○	○	○	○	○			○	○	○
Water																					
• Present cycle: sources, sinks, mechanisms, dynamics?	●						●			●					●				●		
• What is the 3-D crustal water distribution/origin (liquid/ice)?	●						●			●	●	○			○				○		○
• How has the hydrological cycle operated in the past?		●					●				●	○			○						
Life																					
• Does life exist on Mars?	●										●					○	●				
• Can any chemical products of life be detected?	●		●		●		●				●			○	○		●				
• Do isotopic patterns suggest life?							●				●						●				
• What can we learn from Antarctic meteorites?						●															
Atmosphere																					
• What is the current composition of the atmosphere?	●										●				○				○	○	
• What are the circulation dynamics of the atmosphere (T,P)?	●			●			●					●			○				●	○	●
• How has the atmosphere changed over time?											●										
• What is the radiation environment at the surface of Mars?													●				●				
• What is the nature of weather on Mars?	●			●				●			●				●				●	○	●
Climate Change																					
• What is the inter-annual variability of climate?				●							●				○				○		○
• What has been the long-term climate history of the planet?											●										
Upper Atmosphere and Plasma Environment																					
• What are the dynamics of the upper atmosphere?																					
• What are the hot atom abundances and escape fluxes?											●							○	○		
• What are the ion escape fluxes?																		○	○		
• What are the magnetic field configurations?	●			●														○			
• What are the processes controlling the ionospheric energetics?																		○			

MARS SAMPLE RETURNS

Sample-return missions are technologically very challenging. Much must be accomplished in the years between now and the putative earliest date (2011) when the first sample-return mission might be launched. COMPLEX urges that NASA redouble its efforts to develop the essential technologies and infrastructure necessary to make the first sample return a reality in that time. The sample-return strategy is ambitious and exciting, but as currently defined, it depends on numerous technologies and strategies that have not been attempted before. These include precision landing and surface operations, a robust Mars sample collection and containment capability, a Mars ascent vehicle, a strategy for reliable sample recovery and Earth return, and an Earth-based quarantine facility with plans for sample handling and sample distribution to the sample analysis community.

Timing of the First Sample-Return Mission

An important question concerns the timing of sample return. There are two different points of view about when the first sample-return mission should be flown. From one perspective, because the technological and cost requirements are so great for sample-return missions, it is essential for the first sample returned to contain vital information relative to the biological potential of Mars. The other perspective is that a state of diminishing returns has been reached (after the missions through 2005) in acquiring data to identify promising sites; enough is known now to select fruitful sites, and the best strategy is to move to sample return as quickly as possible to guide future Mars exploration. COMPLEX elaborates below on these disparate viewpoints.

The first viewpoint argues that it is likely that the costs of sample return will be high, both in spacecraft resources and in the Earth-based infrastructure to receive and house the samples, and therefore that the number of sample-return missions flown will be small. Sample return should be deferred, therefore, until everything has been done that can be done with remote sensing and through numerous in situ measurements of key indicators such as reduced carbon, to ensure that the samples with the most compelling potential to answer the question, Did life ever arise on Mars?, are obtained. Underlying these arguments, to some extent, is the fear that without the reassurance of exhaustive remote-sensing surveys, the first samples returned may turn out to be indistinguishable from SNC meteorites, and that in such a case, further support of sample return would be jeopardized.

Those with the opposing point of view hold that enough will be known from remote sensing and in situ measurements by 2011 to mount a fruitful mission, and that sample return should be expedited. By the end of the 2005 missions, most of the critical measurements that are needed to guide surface sampling will have been made from orbit. However, the true meaning of the remotely acquired data will not be fully understood because of the lack of ground-truth. Landed science missions will provide some ground-truth, but they are no substitute for the laboratory examination of surface materials. Experience on Earth and on the Moon has shown that significant advances in exploiting remotely sensed data require integration with samples and a knowledge of surface properties. Early sample return will be essential for optimizing future efforts to find the most compelling biological sites. Experience must be gained in the technically complex and risky activity of collecting and returning martian samples, in the process maximizing the value of the remotely acquired data sets for future exploration. Those in this camp argue that it is by the second or third sample-return mission, with attendant periods during which data are evaluated, that the most compelling sites are likely to be revealed and sampled.

COMPLEX emphasizes that answering the question, Did life ever arise on Mars?, must be approached from a broad understanding of the planet and its history. Investigators must be as prepared for an answer of no as well as for yes. It is the committee's view that the goal of the "wait for a perfect sample" viewpoint is too narrow in scope; that approach risks failure if an uninformative sample is in fact returned, and it risks interminable delays if the remotely acquired orbital and in situ data are held by some to be equivocal.

COMPLEX notes that sample return "as soon as possible" will not occur all that soon (≥2011), and in the interim there is time to make additional remote-sensing measurements in support of the first sample return. The committee argues (see Chapter 11) that, with or without additional remote-sensing studies, there is no danger that the surface samples returned by the first mission will be identical to SNC meteorites, or that they will be uninteresting, whether or not they contain evidence bearing directly on the question of martian life.

The committee considers this to be a measured strategy, one that provides an opportunity to learn and to focus on the best sites through experience. It is very possible that studies of the returned samples will reveal new ways of using existing remotely sensed data to seek promising environments for future sampling. COMPLEX fully endorses a 2011 launch date for the first sample-return mission in NASA's plan. Since the study of martian samples in terrestrial laboratories will advance our understanding of Mars to a new level, this date should not be allowed to slip (in the absence of some extraordinary technological problem).

Even if the 2005 missions provide essentially no additional information about possible landing sites, we will have visible images at very high spatial resolution, maps made in both near- and far-infrared wavelengths, and maps of elemental composition. This is enough information to define a half-dozen or more sites that would be excellent starting points for an undoubtedly long campaign of martian field sampling. Moreover, nothing will help with landing-site selection as much as experience gained at the first site, wherever it is. Developments in flight technology can and should continue beyond 2011, but those developments also will be most significantly aided by *experience* with the realities of an active sample-return program.

Successful and productive programs usually involve a series of missions. Flights at the beginning emphasize—and catalyze—technology development. Those later in the series are more productive scientifically both because the technology is better and because the scientific questions and strategies are continually refined. At Mars, a further constructive interaction can be foreseen between a sample-return program and orbital and in situ robotic investigations. (Though the program strategy does not discuss plans for a continuation of orbital and in situ investigations concurrently with sample return, allowance for such missions as new ideas develop should be part of the plan.) Characteristics of the returned samples will suggest objectives and methods for remote-sensing techniques and identify optimal objectives for robotic missions. It will be far better and more realistic to plan for multiple missions and to begin them as early as possible, than to concentrate on attempting to design and manage one or two "perfect" sample-return missions.

Recommendation. Because returned samples will advance Mars science to a new level of understanding, COMPLEX endorses the high priority given to sample return by earlier advisory panels, and it recommends that a sample-return mission be launched at the 2011 opportunity.

How Many Sample-Return Missions Are Required?

Since sample-return missions are expensive, a key question is how many of them are needed. The following factors, partly based on earlier experience with the Apollo program, are important to consider in planning a program of Mars sample return.

Chronology

Measuring the ages of Mars rocks is a crucial goal of the sample-return missions. Ages are needed to calibrate the temporal significance of crater densities on Mars and to attach ages to units in the stratigraphic column of Mars (see Chapter 4), as has been done for the Moon. To accomplish this, more than one site should be sampled in areas representing each of the major martian time periods (Noachian, Hesperian, Amazonian). The need for this duplication of missions lies partly in the uncertainty that a particular mission will be able to sample bedrock and partly in the fact that multiple objectives will be set for each mission, which will make accomplishment of the chronological goal less certain. Ages measured on returned samples are far more precise and can provide more kinds of information than can the in situ age measurements discussed above in this chapter.

Sample Provenances

The association of detached rocks collected at a landing site with the bedrock underfoot may not be straightforward. After two lander missions (Viking and Mars Pathfinder) to Chryse Planitia, uncertainty still remains as to whether the rocks there are from the local substrate or were delivered hundreds or thousands of kilometers laterally by outflow channel floods. Thus, the association of rock ages with ages of the surfaces on which they are collected can be uncertain, and multiple sample sites may be needed to verify the relationship.

Life on Mars

The best environments for extant Mars life are probably below the cryosphere (see Chapter 6), which will be very hard to sample. It will likely be hard or impossible to find live or recently-live organisms in most martian surface environments. However, samples from below the cryosphere may be obtainable in special settings (e.g., among outflow channel deposits), and samples from surface expressions of hydrothermal or aqueous systems may also be obtainable. Intelligent analysis and multiple attempts will be required to collect samples with martian life in them, if there is any.

Technology Development

It is not known how effectively the surface of Mars can be sampled or how readily the rocks collected can be returned. It is worth remembering that the Apollo 7, 8, 9, and 10 missions did little more than test out equipment on the way to the lunar landing. Apollo 11, 12, and 14 were mainly occupied with learning how to sample, move around, and explore. Only the last three Apollos—15, 16, and 17—were serious exploration missions by the criteria of stay-time, experience, crew training, and mobility. Similarly, a certain number of the Mars sample-return missions will be, effectively, technology-development missions.

Increasingly Specialized Needs, Techniques, and Targets

The first Mars sample-return mission should and will employ simple, minimal sample collection techniques. Lessons learned from this mission, and particular sampling needs indicated by the first samples returned (as well as continuing robotic exploration of Mars), will guide the design of increasingly sophisticated follow-on missions. This will repeat the Apollo experience, in which studies of the relatively simply collected Apollo 11 and 12 samples showed the need for more specialized sampling techniques on later missions: for example, when the importance of centimeter-sized lithic fragments in the lunar soil became apparent, sievelike collecting scoops were created for the astronauts to permit fine dust to fall through while retaining the pebbles.

Serendipity

The only thing we can be absolutely sure of finding on Mars is something not expected now. It will be important to keep enough slack in the schedule of sample-return missions to take advantage of the unexpected. For example, if the discovery by Pieters of spectral evidence of olivine in the central peaks of Copernicus on the Moon,[1] signaling the presence there of crater debris from the lunar mantle, had been made a dozen years earlier, one of the Apollo sample-return missions probably would have been sent there. This increases the number of Mars sample-return missions that should be contemplated.

Number of Sample-Return Missions

How many sample-return missions are indicated by these considerations? COMPLEX believes the most useful answer it can supply is that one sample-return mission will not be enough, nor will two or three. The discussion above suggests that on the order of 10 missions may ultimately be required to learn the most important things about Mars, with perhaps three required for initial technology development and shakedown, four focused on planetary chronology, and three on major processes and exobiological sites (not necessarily in that order). Late in the sequence of missions, the committee anticipates that sampling technology may have matured to the point of including a drill capable of reaching depths unaffected by the hostile martian surface environment, and the range of sampling operations may have expanded to include exploration of the polar caps, with the collection and return of ice and dust cores.

Ten is a daunting number of sample-return missions to contemplate. However, after drawing lessons from the Apollo program, COMPLEX wishes to stress an important difference between the Mars Exploration Program and Apollo. Apollo was a "crash program," in which the initial plan was to land 10 manned spacecraft on the Moon in the space of about 4 years. Economy dictated the crowded schedule; it was important to terminate the contracts that provided flight systems and technical support for the missions as soon as possible. COMPLEX is confident the Mars sample-return program will not be executed in this fashion. The committee anticipates, and recommends, that Mars sample-return spacecraft be acquired singly or perhaps in pairs, over a period of years, probably many years. A protracted schedule increases the credibility of the concept of 10 sample-return missions. It will also allow adequate time for the information gained by each mission to be digested and fed into the planning of the next mission, and it will permit substantial redesign of the spacecraft and sampling system between missions. The schedule of the Apollo program and the purchase of multiple copies of the same system greatly limited evolutionary development of this sort.

Recommendation. It should not be anticipated that a few (two to three) Mars sample-return missions will serve the need for samples from that planet. No single site or small number of sites on Mars will answer all of the important questions about the planet, and in any case, the earliest sample-return missions will be in large part technology-development missions. Some 10 sample-return missions, spread over a substantial period of time, may be required to answer the important questions about Mars.

Preparations on Earth for Sample Return

A series of advisory panels have considered the special problems associated with bringing samples from Mars to Earth,[2,3,4,5] and NASA has acknowledged the need to prevent forward and back contamination at every stage of the process of delivery. This includes the need to construct a quarantine facility to receive and contain the samples. NASA's actions to date have consisted of naming a Planetary Protection Officer and sponsoring panel discussions, including those cited, but no concrete action has been taken toward the construction of such a facility.

In a recent report COMPLEX drew attention to the long lead time required to prepare a quarantine facility for the reception of Mars samples once they are delivered to Earth.[6] On the basis of prior experience with terrestrial biocontainment facilities and the Apollo Lunar Receiving Laboratory, the committee estimated that 7 years would be required to design, construct, and staff the facility. To this must be added the time needed to clear an environmental impact statement and to carry out several reconnaissance studies that are needed to inform the design and operation of the facility.[7] The aggregate of time required will strain the schedule even of a 2011 launch (with a 2014 return). The message is plain: preparations for a sample return should not be delayed any longer than they already have been.

Recommendation. Scientific research and design studies that must precede the design and construction of the Mars Quarantine Facility should begin immediately. Decisions should be made immediately about the siting and management of the facility. Design and construction of the facility should begin at the earliest possible time. (This recommendation also appears in Chapter 7.)

OTHER ISSUES RELATING TO MARS EXPLORATION

Power Supply for Landers and Rovers

An extremely important consideration in establishing the capabilities of landed packages, static or roving, on Mars is the power supply on which they rely—the options being solar panels and radioisotope power systems. This becomes important with the Mars Science Laboratory, an advanced rover scheduled for launch in 2007,[d] and later landers, including sample-return missions.

The Viking landers operated for 7 years; they were able to do so because they were equipped with radioisotope power systems. On the other hand, the Mars Exploration Rovers (MERs) to be launched in June 2003 are equipped with solar panels, and are not designed to operate for longer than an estimated 90 days. The lifetime is limited because as the maximum elevation of the Sun in the martian sky declines, the available solar power decreases; for the same reason, the rovers get colder and need more power to keep warm. Meanwhile, dust accumulates on the panels, further reducing the power. The MERs are also constrained by the needs of their solar panels to land in the 10°N to 15°S latitude belt. Because of the 90-day time limitation, the MERs will be able to stop and make measurements on rocks no more than six times. These are very severe limitations on the utility of the MER system.

The situation for MSL is similar if it is powered by solar panels. A possible solution would seem to be to increase the size of the solar array. This would be feasible on a stationary lander, but it would be problematic on a rover, which the MSL is, because solar panels that overhang the sides of the vehicle would be vulnerable to damage as the vehicle drove over a rocky surface. Thus, the MSL also will have a limited lifetime and range if it is solar-powered. A drill has been projected for the MSL, to collect samples from up to 2 m in depth; but drills are power-hungry, and they need time. Appendix A (see the subsection "In Situ Analysis and Returning Samples to Earth") acknowledges that the use of a drill requires "reliable, long-lived power." The other functions of MSL also require better than solar power.

The power problem could have a serious impact on a sample-return mission. Reliance on solar power would mean that samples would almost certainly have to be collected at low latitudes, which excludes those parts of Mars where ground ice is stable and where other volatiles are most likely to be present. Again, if the sample-return mission has a rover to collect samples, its lifetime will be short. This, coupled with the slowness of moving across the martian surface and the necessity of looping back to the lander to deliver samples, will restrict the region of sampling to very close to the lander, probably within a radius of a few hundred meters. An option might be to acquire the samples by drilling for a long time from a stationary lander with a very large solar array.

The message is that the more sophisticated stages of Mars surface exploration will be severely limited if they are forced to rely on solar power, and it seems essential that these missions be equipped with radioisotope power systems. Radioisotope power systems have been considered off-limits for the last few years because the pertinent nuclide, ^{238}Pu, has not been produced in the United States during that time. However, a record of decision dated January 19, 2001, on a programmatic environmental impact statement concerning Department of Energy nuclear facilities, stated the intent of the Department of Energy to reestablish a domestic capability to produce ^{238}Pu for future space missions. That record of decision also allowed for interim purchase of Russian ^{238}Pu, if required.

COMPLEX understands that even if they are available, radioisotope power systems are expensive. However, from a science viewpoint the advantages of nuclear power—long-lived missions and access to any point on Mars—are clear. COMPLEX urges the use of nuclear power sources, if at all feasible, on advanced Mars lander missions.

The Scout Program

The Mars Scout program provides an excellent opportunity for NASA to accommodate science topics outside the themes of water and life that are the focus of Mars missions. If the Scout program is to be modeled after the successful Discovery program, it is essential that the science goals for the Mars Scout program be directed toward the highest-priority science for Mars, and not only toward the themes of water and life.

There is concern in the Mars science community that the Scout program, the youngest and smallest element of the Mars Exploration Program, may also be the most vulnerable. The fear is that the Scout program may not achieve its potential because it will be sacrificed in times of budget stringency. COMPLEX considers that this would be unfortunate.

---

[d]Following the completion of this study, NASA announced that it was delaying the launch of the Mars Science Laboratory until 2009 to allow time to develop an advanced, radioisotope power system for this mission.

Recommendation. So that the Scout missions can fulfill their laudable goals of filling in gaps in the Mars Exploration Program and allowing a rapid response to scientific discoveries, COMPLEX recommends that care be taken to maintain this program as a viable line of missions when budget problems arise.

Data Analysis, Ground-Based Observations, and Laboratory Analysis

The Mars Exploration Program, with its missions at 2-year intervals, presents a new problem in fully exploiting the amount and variety of data that will be collected. The volume and quality of data returned by MGS alone have been extraordinary, and the analysis of these data is only beginning. With the rapid pace of Mars missions planned for the next decade, the flood of data can be expected to increase. This problem should be recognized, and NASA's data analysis and science programs should be structured to accommodate and support the broad range of Mars science that is to come.

While the Mars Exploration Program consists of flight missions, exploration and understanding of the planet as a system also depends upon other modes of data acquisition, including ground-based and Earth-orbital observations, antarctic meteorite studies, laboratory analysis, and theoretical modeling (see Chapter 1). These are all essential components of Mars science.

Recommendation. A plan should be developed at the program level, not at the level of each mission, for archiving and making accessible the data to be gathered by the Mars Exploration Program. It is essential that support be provided for the study and exploitation of this body of data.

Recommendation. COMPLEX endorses continued support for nonflight activities such as ground-based observing and laboratory analysis.

REFERENCES

1. C.M. Pieters, "Copernicus Crater Central Peak: Lunar Mountain of Unique Composition," *Science* 215: 59–61, 1982.
2. NASA Astrobiology Institute, *Mars Program Architecture: Recommendations of the NASA Astrobiology Institute*, Ames Research Center, Moffett Field, Calif., 2000.
3. NASA, *Mars Sample Handling and Requirements Panel (MSHARP) Final Report*, NASA/TM-1999-209145, Washington, D.C., 1999.
4. Space Studies Board, National Research Council, *Mars Sample Return: Issues and Recommendations*, National Academy Press, Washington D.C., 1997.
5. NASA, *Mars Sample Quarantine Protocol Workshop: Proceedings of a Workshop Held at NASA Ames Research Center, June 4–6, 1997*, NASA/CP-1999-208772, Washington, D.C., 1999.
6. Space Studies Board, National Research Council, *The Quarantine and Certification of Martian Samples*, National Academy Press, Washington, D.C., 2002.
7. Space Studies Board, National Research Council, *The Quarantine and Certification of Martian Samples*, National Academy Press, Washington, D.C., 2002, p. 2.

13

Conclusions

It is humankind's nature to explore our surroundings if it can be done. Fifty years ago, exploring Mars was not one of the things anyone could do. Those who were curious had to be content with fuzzy images of the planet, quivering in the oculars of telescopes. But that is far from the case today. Forty years ago, spacecraft began to be sent to the planets, and since then, the art of space exploration has become increasingly refined and discoveries have multiplied. We now have the capability, in principle, of reaching and exploring any object in the solar system. At the top of the list of targets of exploration is Mars, the most Earth-like, most accessible, most hospitable, and most intriguing of the planets. Two years ago, in October 2000, NASA recognized this by setting the study of Mars apart in a structured Mars Exploration Program. The present document reports on COMPLEX's study of the program.

COMPLEX has compared the elements of the Mars Exploration Program with the research objectives for Mars that have been stressed by advisory panels, including this one, for more than 23 years. The committee found that correspondence between the two is not perfect. Currently, NASA focuses on the search for life, and its prerequisite, water, as the main drivers for Mars research, and has favored missions and experiments that support these goals. The space agency is not now in a position to ask direct questions about life on Mars, and has not been since the Viking mission in the 1970s, but the missions supported are designed to find the areas most promising for water and life, and to investigate in situ their chemical and petrographic potential for extant or fossil life.

Since NASA operates within budget constraints, this emphasis on one particular scientific objective necessarily comes at the expense of others. COMPLEX considered the question of whether NASA's priorities are too heavily skewed toward life-related investigations. The committee decided, however, that this is not the case. The emphasis on life is well justified; the life-related investigations that are planned range over so much of Mars science that they will result in broad and comprehensive gains in our knowledge; and the areas most neglected as a consequence of this emphasis (see Chapter 12) will, to some extent, be investigated by projected missions of our international partners.

COMPLEX endorses the program NASA has set up, though the committee has also pointed out several areas of high scientific priority that the program does not address. This report stresses the uniquely important role of sample return in a program of Mars research, and urges that sample-return missions be performed as early as possible. Discussions and recommendations related to sample return appear in Chapters 7 and 12. A more general review of the conclusions of this report is contained in the Executive Summary.

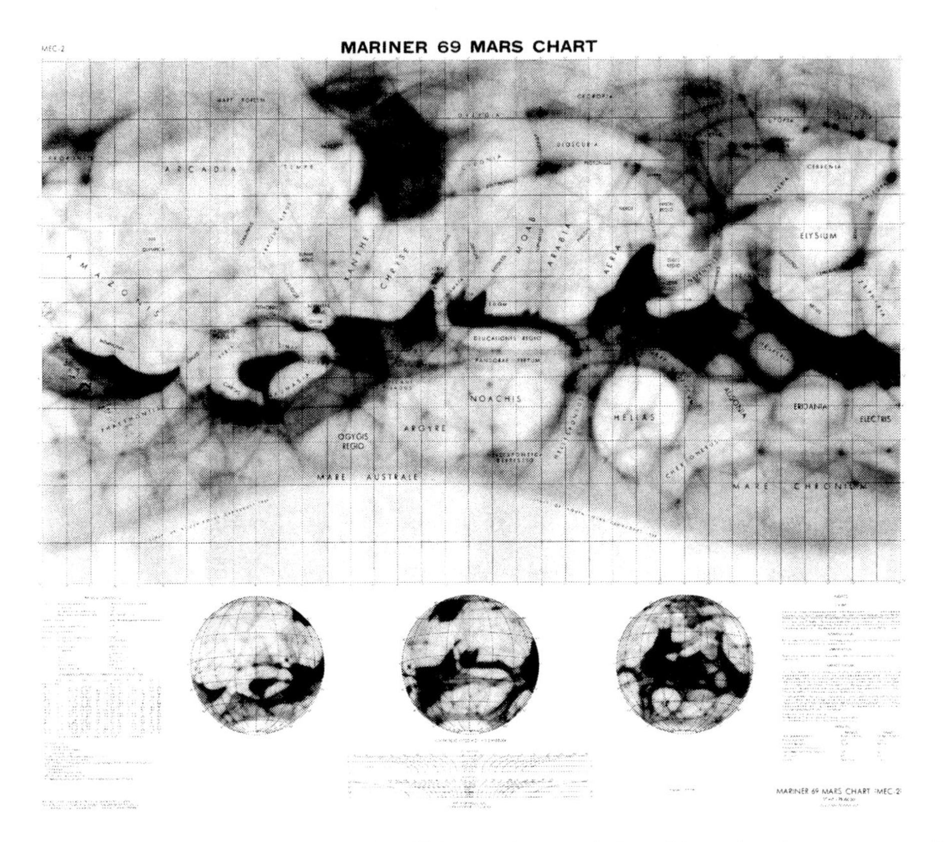

FIGURE 13.1 The study of Mars has come very far. This map is a reminder of how the planet was perceived in 1967. SOURCE: Mariner 69 Mars Chart, NASA MEC-2.

Our understanding of the most Earth-like planet beyond our own has increased dramatically in 35 years of spacecraft research (see Figure 13.1). Most of us will live to see an even greater increment of knowledge result from execution of the Mars Exploration Program that this report describes.

Appendixes

Appendix A

The NASA Mars Exploration Program

A summary of active and planned missions of NASA's Mars Exploration Program is presented in Table A.1 pn pages 110 and 111. Foreign as well as NASA missions are included so that the totality of Mars research can be evaluated. The section that follows on pages 112 through 117, written by Orlando Figueroa, director, and James B. Garvin, lead scientist, of the Mars Exploration Program, Office of Space Science, NASA Headquarters, enlarges upon details of the missions and the overall strategy of the program. It should be stressed that the presence of an integrated Mars program at NASA, as opposed to a series of isolated missions to Mars, is a relatively new development.

TABLE A.1 Active and Planned Missions to Mars, by NASA and Foreign Space Agencies

Launch Date	Mission	Instrument Complement	Science Objectives and Other Comments
1996	NASA Mars Global Surveyor (MGS)	Mars Orbiter Camera Thermal Emission Spectrometer Mars Orbital Laser Altimeter Radio Science Magnetic Fields Investigation	Recovering some of the Mars Observer (MO) objectives; acquiring systematic global data sets. High-resolution imaging. One orbiting spacecraft, polar orbit.
1998	NASDA (Japan) Nozomi	Mars Imaging Camera Neutral Mass Spectrometer Thermal Plasma Analyzer Mars Dust Counter Radio Science Experiment Plasma Waves and Sounder Low Frequency Plasma Wave Analyzer Ion Mass Imager Magnetic Field Investigation Probes for Electron Temperature Measurements Ultraviolet Imaging Photometer Electron Spectrum Analyzer Energetic Ion Spectrometer Extreme Ultraviolet Spectrometer	Space physics; motion and structure of upper atmosphere and ionosphere. Intended to arrive at Mars in 1999, but technical problems have delayed arrival until 2004. One spacecraft in elliptical orbit.
2001	NASA Mars Odyssey Orbiter (MO 2001)	Mars Radiation Environment Experiment Thermal Emission Imaging System Gamma Ray Spectrometer	Recovering some of the Mars Observer objectives; systematic global measurement of chemical elements, H, and minerals. One orbiting spacecraft, polar orbit.
2003	NASA Mars Exploration Rover (MER)	Panoramic Camera Miniature Thermal Infrared Spectrometer Microscopic Camera Mössbauer Spectrometer Alpha Particle-X-Ray Spectrometer	Determining the aqueous, climatic, and geologic history of two sites where conditions may have been favorable for the preservation of biotic or prebiotic processes. Two landed rovers.
2003	ESA Mars Express (ME)	High Resolution Stereo Color Imager IR Mineralogical Mapping Spectrometer Planetary Fourier Spectrometer UV and IR Atmospheric Spectrometer Energetic Neutral Atoms Analyzer Subsurface Sounding Radar/Altimeter Radio Science Experiment	Global mineralogic and high-resolution stereo mapping, subsurface sounding, atmospheric dynamics. One orbiting spacecraft, polar elliptical orbit.
2003	U.K. Beagle 2	Gas Chromatograph-Mass Spectrometer Microscope Camera Panoramic and Wide Angle Cameras Mössbauer Spectrometer Alpha Proton X-Ray Spectrometer Environmental Package Sensors	Investigating the geology, geochemistry, and exobiology of a landing site. One lander.
2005	NASA Mars Reconnaissance Orbiter (MRO)	Recommended: Pressure-Modulator Infrared Radiometer Mars Color Imager Wide Angle Camera Mars Color Imager Medium Angle Camera Visible near-IR imaging spectrometer Visible imaging camera Radar sounder Radio science investigations	Recovering the MO and MCO atmosphere and climate science objectives; searching for sites showing evidence of aqueous and/or hydrothermal activity; exploring in detail hundreds of targeted, globally distributed sites; searching for liquid or frozen water in the near surface. One orbiting spacecraft, polar orbit.

TABLE A.1 Continued

Launch Date	Mission	Instrument Complement	Science Objectives and Other Comments
2007	ASI (Italy)/ NASA Orbiter	(To be determined)	Telecommunication relay
2007	Orbiter, CNES	(To be determined)	Prototype orbiter; orbital science and NetLander transport, demonstration of aerocapture and orbital rendezvous, telecommunication relay for NetLander mission. One orbiting spacecraft.
2007	NetLander, European consortium (CNES, DLR, FMI, ASI, Belgium and others)	Some of these experiments: Surface Meteorology Package Electric Field Ground Penetrating Radar Magnetometer NetLander Ionosphere and Geodesy Experiment (plus Total Electron Content) Panoramic Camera Seismometer Soil Properties, Thermal Inertia and Cohesion Experiment Mars Microphone	Objective is to perform simultaneous measurements in order to study the internal structure of Mars, its subsurface and its atmosphere. Network of four stationary landers.
2007	NASA Mars Scout	(To be determined)	Discovery-class competed mission opportunities, $300 million cost cap, PI-led; expected to meet science goals and opportunities not covered by other missions.
2007	NASA Mars Science Laboratory (MSL)	(To be determined)	Long-range, long-duration advanced science laboratory. A pathfinding mission for sample return to demonstrate "smart landers" technology, including accurate landing and hazard avoidance. One landed rover. Delayed to 2009 to allow for the inclusion of an advanced radioisotope power system.
2009	ASI (Italy)/ NASA Orbiter	(To be determined)	Possible synthetic aperture radar system.
2011	NASA Mars Sample-Return mission (MSR)	(To be determined)	

THE MARS EXPLORATION PROGRAM: A HIGH-LEVEL DESCRIPTION[a]

James B. Garvin and Orlando Figueroa
Office of Space Science, NASA Headquarters

The newly restructured Mars Exploration Program (MEP) is fundamentally a science-driven program whose focus is on understanding Mars as a dynamic "system" and ultimately addressing whether life is or was ever a part of that system. It further embraces the challenges associated with the development of a predictive capability for martian climate and how the role of water, obliquity variations and other factors may have influenced the environmental history of Mars. This white paper outlines in a high-level sense the scientific strategy of the new Mars Exploration Program. As the MEP will continuously evolve in the context of the scientific discoveries achieved and the changing character of the scientific drivers provided by the broad scientific community to NASA, it is important to recognize that the present strategy is a living one. The foundation of the present strategy is often referred to as "Follow the water" and this serves to connect fundamental program goals pertaining to biological potential, climate, the evolution of the solid planet, and preparations for eventual human exploration. Balance is recognized as important within the present MEP, given the challenges associated with assessing the biological potential and climate record of a distant object such as Mars.

On the basis of the knowledge gained from the Mariner, Viking, and Mars Global Surveyor missions, we know that Mars, like Earth, has experienced dynamic interactions between its atmosphere, surface, and interior that are, at least in part, related to water. As NASA embarks upon an intensive program of scientific exploration of Mars, following the pathways and cycles of water has emerged as a strategy that may lead to a possibly preserved ancient record of biological processes, as well as to an understanding of the character of paleoenvironments on Mars. In humanity's exploration of extreme environments on Earth (the deep ocean, ice fields, geothermal sites, and so on), wherever there is liquid water below the boiling point, evidence of life has been identified. The presence of liquid water sometime and somewhere in the martian past, coupled with other key variables (temperature, pressure, soil chemistry, and atmospheric chemistry) makes Mars an attractive target in expanding the scientific understanding of life, its origins, and diversity within the Universe. In addition, other scientific drivers have emerged, including the use of Mars as a location from which to provide absolute calibration of the timing of major Solar System events.

One example of the difficulties associated with understanding a planet as complex as Mars is associated with the assessment of its biological potential. Searching for evidence of existing or ancient life on Mars is fraught with seemingly insurmountable challenges. There is a tremendous surface area within which to search—150 million square kilometers of martian surface, roughly equivalent to the continental land mass of Earth. In addition, even after five major missions to Mars, comparatively little is known about the characteristics of the upper surface layer, and of the impact of ultraviolet and cosmic radiation upon the surface environment. Thus, evidence of potential life may lie tens or even hundreds of meters below within the naturally shielded shallow subsurface. Our existing knowledge of the martian shallow subsurface remains purely inferential, yet predictive models indicate that liquid water could be stored in subsurface reservoirs. Such environments may be compelling localities for in situ exploration.

The Viking missions of the 1970s searched for answers to the "life question" by directly seeking evidence of biological activity in the upper 10 cm of the surface at two widely separated landing sites. In effect, the Vikings sought to hit a scientific "home run" on our first attempt at bat. The life detection experiments on Viking were arguably the best available, given 1970's technology and the limited understanding of *how* to detect life in general in extreme environments. Unfortunately, the results of the Viking in situ life detection experiments were inconclusive. Given our then limited knowledge of Mars, the two Viking landing sites selected were reasonable starting places, but we now know that Mars is a remarkably diverse and dynamic planet, with many distinct regions that may differ significantly in their potential for harboring records of existing or ancient life. In retrospect, in spite of the bold approach adopted by the Vikings in the 1970s, Mars was not quick to yield its secrets, and in the ensuing 25 years NASA and the scientific community have developed an improved framework for examining Mars, just as has been done for Earth. What we have found is a planet exceedingly rich in landscape diversity, with what appears to be a preserved record of sediments that may be an indication of a major role of liquid water in its earliest epoch.

[a]The program description, a "living document" that is updated periodically, is reprinted here as received at the time of the study.

As an example, the search for life or life-generating environments on Mars requires a systematic approach through which we can begin to understand the complex systems of geology, climate, and biological potential that constitute the "Mars System." In order to understand Mars as a dynamic system, we must first establish a global context of information about the planet, and then validate and expand that knowledge by increasingly narrowing our focus through surface investigations, ground-truthing, and targeted reconnaissance. With a strong foundation of orbital and surface reconnaissance and directed investigations, we can then make a well-informed selection of the most-promising local sites from which to obtain samples for return to Earth for comprehensive analysis. We refer to this approach as "Seek, In-situ, Sample"—increasingly narrowed cycles of "seeking," first from orbit, then on the surface, followed by collection of well-selected samples for return to Earth.

This approach parallels that used in exploration for minerals and other natural resources here on Earth. Petrochemical companies use satellite imagery of Earth to identify regions where there are chemical indicators of processes that concentrate valuable materials in certain geologic settings. They then follow up with localized analysis of those regions, before sending in the "wildcat" crews to drill for resources beneath the surface. The difference in our exploration of Mars is that rather than prospecting for oil or natural gas, we are prospecting for water, signatures of life, or ancient environments conducive to life as we currently understand it.

In attempting to understand the "real" Mars, it is first necessary to inventory the key constituents of the Mars system. This step is needed to establish the foundation or context within which particularly difficult questions (i.e., such as "Was there ever life on Mars?") can be addressed.

Thanks to Viking and the ongoing Mars Global Surveyor (MGS) mission, we now know Mars has experienced wide swings in its climate, potentially extensive periods in its past when liquid water was persistent at the surface in localized depressions, and that it may harbor an active subsurface hydrological system even today. In spite of the failures of the Viking surface laboratories to detect any signs of past or present life in the 1970s, we can refine the approach and continue the quest with real prospects of making major strides during the current decade.

Global Mapping—The Foundation for Context

How will we attack the mysteries of Mars in order to determine whether it ever harbored life or had experienced climate oscillations that mimic those of Earth? These questions have puzzled planetary scientists for 25 years and today, thanks to the global mapping of the MGS, a refined and robust strategy has emerged.

For example, in order to most aggressively address whether life ever existed within the Mars system, it is first necessary to have a systematically increasing body of knowledge about the global surface, atmosphere, hydrosphere (although it may be entirely frozen into a cryosphere), and interior. A comprehensive global inventory of the martian surface (and shallow interior) then enables detection of localized "anomalies"—places that are different in terms of chemistry, temperature, venting of important gases, or other factors. This "prospecting" step is a key initial part of our strategy. We must *seek* the most promising places on the surface of Mars to continue intensive local reconnaissance in the context of a global picture of Mars.

Thus, the first step in our Mars exploration strategy is to acquire adequate global reconnaissance using orbital remote sensing tools that not only define Mars in a global context but also tell us where to look in our refined search for localized "hot spots"—places where the action of liquid water and possibly temperature have provided "fingerprints" that we can identify from orbit. The MGS mission currently mapping Mars in combination with the Mars Odyssey orbiter constitutes the first wave of systematic orbital reconnaissance of Mars. The aim is to continuously refine our search for areas on Mars where higher-resolution and landed investigations can best continue the exploration for those materials that offer the most promising prospects for resolving issues related to life, timing of events, and climate history.

At issue is what percentage of Mars today might be identifiable as "hot spots." Recent findings from the imaging systems and spectrometers aboard the Mars Global Surveyor suggest that there may be a few hundred to perhaps a thousand "hot spots" worthy of near-term intensive investigation at the surface. Isolating the handful of the most scientifically compelling poses a challenge to the reconnaissance elements of our unfolding Mars Exploration Program.

Data from Viking and MGS suggest that certain landscapes on Mars were likely sculpted by the action of liquid water. However, those same formations might also have been formed as a consequence of exotic processes associated with wind, or even explosive volcanism. We cannot discriminate between the possibilities until we can go to those identified regions and study them at "sample scales"—both at the appropriate scale for definitive process identification, as well as at microscopic scales for provenance studies. Therefore, the next step in refining our global understanding of Mars from this first wave of orbital reconnaissance is to conduct surface investigations at some of the most scientifically compelling sites. Such surface-based investigations will validate and calibrate our global remote sensing data and demonstrate that the mapping from orbit matches the reality of the surface. Data from MGS and Odyssey will identify hundreds to thousands of the most promising regions for these intensive surface investigations.

Surface Investigation and Ground-Truthing

The 2003 Mars Exploration Rovers (MERs) will take those next steps in making the discoveries that could lead in the future to determining whether or not life ever arose on Mars. The mission of the MERs is to find conclusive evidence of water-affected materials on the surface. They are designed to effectively serve as robotic field geologists, and they will provide the first microscopic study of rocks and soils on Mars. The Mars Pathfinder Sojourner rover analyzed eight rocks and soil patches with one inadequately calibrated instrument (APXS) in 83 days of surface operation. The twin MERs will study dozens of rocks with at least three different calibrated instruments, as well as capturing spectacular context images together with mineralogy (from hyperspectral middle-IR imaging). The twin MERs will also have the mobility to wander up to 1,000 meters across the martian landscape, measuring the chemical character of the soils, rocks, and even the previously inaccessible interiors of rocks where unaltered materials may lurk. Just as human field geologists study Earth by using a hammer to break open rocks, the MERs will employ a rock abrasion tool to scratch beneath the outer covering of Mars rocks and look inside with microscopic resolution. Evidence from martian meteorites indicates that carbonates existed on Mars at one time in its past, at least at microscopic scales. What we don't know is whether carbonates existed at or near the surface, or whether they were produced in association with biological processes. The MER robotic geologists will help answer this question. Finally, the MER perspective will allow for quantified calibration and validation of orbital remote sensing data at the surface that will hopefully yield the capacity to extrapolate to other places on Mars that are similar to the MER landing sites.

By studying the rocks and soils in a "hot spot" region chosen from the MGS and Odyssey reconnaissance data, the MERs will tell us whether what we see from orbit is what we anticipate, and if not, what it may represent instead at least chemically. The MERs will link the surface chemistry and mineralogy at the surface with that surmised from orbital observations and facilitate extrapolation across many different places on Mars.

Armed with this new knowledge and characterization of water-related geologic regions on Mars, we are ready to climb to the next level of reconnaissance and in-situ investigations—including the search for life-bearing environments within those water-related regions as well as the surface record of climate.

Targeted Reconnaissance and Landing Site Characterization

The 2005 Mars Reconnaissance Orbiter (MRO) is the ultimate reconnaissance tool in the Seek, In Situ, Sample strategy. Following the surface validation and investigations of the 2003 MER twin rovers, the MRO will narrow the focus into the localities identified from MGS and Odyssey to search for the most compelling environmental indicators suitable for bearing life (warm, wet, chemically benign, etc.) or recording aspects of climate. The MRO will use its new observational tools, some of which could resolve beachball-sized objects and their mineralogies, to search for clues within the martian landscape of telltale layers and materials associated with the action of liquid water.

Recent evidence suggests that water-related mineral indicators may be detectable from orbit at certain specific infrared wavelengths provided high enough spatial resolution is adopted. While debate lingers within the science community about resolution thresholds, imaging thousands of promising sites at ~30-cm resolution would allow discrimination of water-related sedimentation from that associated with explosive volcanism (layers of cemented ash), wind, or global dust settling. Thus, the MRO seeks to develop a globally-distributed set of panchromatic and hyperspectral images that isolate the dozen or so most compelling sites for intensive surface-based exploration and sample return. When coupled with the first in-situ examination of two different water-related sites on Mars as provided by our twin MERs, our approach allows us to build confidence we can predict how Mars operates under certain conditions, as well as to demonstrate the past action of water in the otherwise hyper-arid desert that characterizes the modern Mars of today. In addition, the MRO will seek to develop the first understanding of modern water as it behaves within the present martian atmosphere and how climatology operates on annual basis. The MRO will also attempt to characterize the shallow subsurface of Mars in search of water-related layers or deposits and other stratigraphic indicators of ancient water-related environments.

While MGS and Odyssey may identify hundreds to thousands of interesting places on Mars, only two can be visited and evaluated by the twin MERs, in part because of their landing precision (50 km at best). The MRO is designed to evaluate the most compelling places identified previously and to measure their prospects at new scales, wavelengths, and with tools that measure the shallow subsurface vertical structure. The "hottest" places for intensive surface reconnaissance and exploration identified from MGS and Odyssey will be exhaustively targeted by MRO so that by 2006, a set of compelling localities will be established within a global scientific framework. Of these, the top two or three in terms of their potential as martian biomarker sites will serve as the bridge into the second phase of our strategy. Of course, scientifically compelling sites may include localities where the climate record is best exposed in a physical and chemical sense.

In Situ Analysis and Returning Samples to Earth

The final part of phase one of our Mars Exploration strategy begins after MRO has "fingered" where to go. If all goes as planned, in 2007[b] we will launch a precision landed mobile surface laboratory—the Mars Science Laboratory (MSL)—to the most promising of the targeted sites that will operate on the martian surface for at least half a year. Planned development of a new suite of miniature analytical instruments for this mobile laboratory, which are tuned to questions of geochemistry and biological processes, will measure aspects of the surface and subsurface materials potentially linked with ancient life and climate. Plans include a laser Raman spectrometer to focus our surface search for carbonates, a micron-resolution optical microscope to assess patterns of micro-scale features, and an instrumented drill that could reach 2 to 3 m beneath the surface in search of buried ice or other "shielded" substances. MSL—also known as the Mobile Science Laboratory or the Mars Smart Lander—could also explore the shallow (~100-m) subsurface using ground penetrating radar or other electromagnetic sounding approaches. With some cooperation from Mars, the MSL could confirm the surface presence of water-related minerals and carbonates and their formational histories. MSL will also have the benefit of the investment in continuous refinement of how orbital remote sensing can be used as a pathfinder to those surface localities that offer the highest probability of harboring martian "fossils" or other forms of indicators of past life. It will serve as both a scientific and technological pathfinder for the robotic sample return campaign that forms the ultimate step in our Mars Exploration Program.

This phase of in situ analysis will incorporate technological advances that permit mobile surface laboratories to be landed within a few kilometers or less of any interesting spot on Mars. By precision landing near to a telling site and having the longevity and mobility to explore as if we (humans) were there, we can extend our search for life-signs and other scientific indicators on Mars to horizontal scales on the surface that will be measured in multiple kilometers, and not football fields. MSL will serve as the bridge to the next phase of Mars exploration: a future series of missions that will endeavor to bring preserved samples of the most interesting materials back to Earth, in context, and with real prospects for harboring bio-signs or chemical indicators of warmer and wetter past environments. Clearly a campaign of sample returns will be needed to tie major Mars system events to the absolute chronology of the solar system. When Mars may have been more biologically hospitable and what the global planetary state of evolution was at that time are vital if understanding is to be achieved.

Thus, our refined strategy seeks to establish a suite of the most promising places for intensive surface analysis prior to the technological leap of returning samples to Earth for analysis. Once we have identified the hottest prospects, a program of long-duration, and reasonably long-range mobile surface laboratory(ies) must be sent to Mars to unravel what is in the rocks, soils, ices, and atmospheric constituents that could be linked to favorable environments for biology and for its preservation or other driving questions. Under this strategy, with good fortune, by the end of

[b]Editor's Note: Following the completion of this study, NASA announced that it was delaying the launch of MSL until 2009 to allow time to develop an advanced, radioisotope power system for this mission.

2008 we could be receiving images of martian microscopic features not unlike those identified by D. McKay and E. Gibson from the Allan Hills meteorite, ALH84001.

The overarching science thrust of the Mars Exploration Program is to examine the diversity of Mars by investigating multiple sites with mobile surface laboratories. However, this would require additional budget resources; a more aggressive program with additional budget increments would allow the development of more than one mobile surface lab and enable the exploration of multiple "hot spots" identified by the MRO and its forerunners. It would also facilitate the earliest possible implementation of sample-return missions. In addition, with the addition of reliable, long-lived power, it could enable direct access to regions up to 20 m deep within the subsurface. One means of providing balance, innovation, and adaptability in our MEP is the Mars Scout Program. This program, currently planned for inception with the 2007 launch opportunity, will solicit principal-investigator-led missions to explore Mars in focused ways not currently baselined in the core MEP program of flight missions. The aim is to promote scientific innovation within the MEP by allowing the broad community to compete for a "Discovery class" mission every other launch opportunity. Given the somewhat limited breadth of the core MEP flight program, and the recommendations of the Mars Exploration Payload Assessment Group (MEPAG), there is plenty of room for additional, high-science-value missions, within the overall Mars Exploration Program. The 2007 Scout competition is expected to engage the Mars scientific and engineering community and deliver an innovative mission with a specific focus not emphasized or treated within the core program. It is possible that the first Mars Scout Mission will involve an array of small surface stations to explore the surface diversity of the planet, or an orbiter to map aspects of the global Mars that cannot be implemented on the 2005 MRO mission. A preliminary set of 10 Scout Mission concepts were selected for 6-month study as of June of 2001.

Finally, the first of several "informed" Mars sample-return missions (MSRs) is planned for a late 2011 launch, with return of ~1 kg of sample materials by 2014. The specific scientific scope of the first MSR mission remains in the hands of science definition teams, but the intent is to build upon the technologies utilized in 2007 for the mart mobile surface laboratory to selectively screen samples at a surface site selected to provide the best sedimentary record of water-related materials indicative of perhaps hospitable paleo-environments. Given the material diversity of Mars and the challenges presented by sampling one scientifically-compelling locality to provide definitive answers to the driving scientific questions about Mars, it is unlikely that a single MSR mission to a sedimentary site will fulfill the scientific requirements and needs. Thus, NASA plans to implement a campaign of MSR missions in the next decade (2011–2020) that will involve long-lived surface exploration with subsurface access to provide the most diagnostic materials for analytical investigations in terrestrial laboratories.

Summary

Finding the right places to go on Mars's vast surface to pursue the search for origins of life, the record of climate, and ultimately our place as humans within the cosmos is the first step required in the new Mars exploration strategy. A natural extension of this progressive strategy of orbital, surface, and ultimately sample-based reconnaissance is to visit two or more sites with precision-landed mobile surface laboratories by the end of the decade. Such a strategy, when combined with small, totally competed missions in 2007 and beyond, will extend our ability to search for elusive clues to the possibilities of life or at least for evidence of ancient, warm, wet environments. Ultimately the surface-based search, in the context of orbital "foundation" data sets, will yield Mars's secrets and allow us to return samples safely to Earth for unprecedented analytical scrutiny.

The new Mars Exploration Program delivers a continuously refined view of Mars with the excitement of discovery at every step. What might we find as we move along the roadmap of mission events this decade? MGS has already discovered possible clues to a source, however small, of modern liquid water. Odyssey could discover carbonates at the surface or regions of enrichment in hydrogen, as well as evidence of possible martian geothermal vents. The twin MERs may discover local evidence for how water once persisted at the surface and what ultimately to search for from orbit. The MRO may discover ancient martian "oases"—localities where chemical and morphological evidence of past warmer and wetter environments is preserved. Alternately, MRO could uncover the shallow subsurface of Mars and voluminous repositories of buried water ice. In 2008, the first of potentially several mobile surface laboratories could find specific materials indicative of locally warm, wet paleo-environments and possibly the first in situ detection of martian organics. A Mars Scout mission in 2007 could sample martian atmospheric dust or probe the workings of martian meteorology. Ultimately the first martian samples of rocks, soils, dust, and perhaps volatiles, will arrive on Earth by 2014, isolated to preclude contamination. These first samples, collected in careful context, will open the door for human missions to Mars sometime in the future.

NASA has fashioned a strategy that is risk attentive, including a natural responsivity to science challenges that will emerge as discoveries are made. It is linked to our experience exploring the deep ocean here on Earth, as well as part of a strategy that uses Mars as a natural laboratory for understanding life and climate on Earth-like planets other than our own.

Appendix B
Compilation of Recommendations Concerning Mars Exploration Made by COMPLEX and Other Advisory Groups

The recommendations reprinted in this appendix are sorted by sources and ordered by date.

1. 1978

Strategy for Exploration of the Inner Planets: 1977–1987 (NRC, COMPLEX, 1978)

[1.1] A global map or image of the surface of a planet at good resolution is considered to be a major scientific contribution and is basic to any advance in the understanding of the terrestrial planets. [p. 42]

[1.2] Two important precepts have guided the Committee's definition of primary objectives for future exploration of Mars. First is the need to carry out intensive studies of the chemical and isotopic composition and physical state of martian material to determine the major surface-forming processes and their time scales and the past and present biological potential of the martian environment. Second is the need to achieve a broad-based and balanced planetological characterization in order that meaningful comparisons can be drawn between Mars and the other members of the triad Earth-Mars-Venus. [p. 43]

[1.3] . . . [t]he primary objectives in order of scientific priority for the continued exploration of Mars are. . . the intensive study of local areas (a) to establish the chemical, mineralogical, and petrological character of different components of the surface material, representative of the known diversity of the planet; (b) to establish the nature and chronology of the major surface forming processes; (c) to determine the distribution, abundance, and sources and sinks of volatile materials, including an assessment of the biological potential of the martian environment, now and during past epochs; (d) to establish the interaction of the surface material with the atmosphere and its radiation environment. . . .

These objectives are multiply connected. For example, definition of the volatile inventory should pay proper attention to gas exchange between the planet and the solar wind.

In the following we will briefly expand on the substance of these recommended objectives and outline the recommended strategy for accomplishing them. [p. 44]

[1.4] The establishment of the chemical, mineralogical, and petrological character of the various components of the martian surface material should include (in approximate order):

1. Gross chemical analysis (all principal chemical elements with a sensitivity of 0.1 percent by atom and an accuracy of at least 0.5 atom percent for the major constituents).

2. Identification of the principal mineral phases present (i.e., those making up at least 90 percent of the material in soils and rocks).

3. Establish a classification of rocks (igneous, sedimentary, and metamorphic) and fines that define martian petrogenetic processes.

4. State of oxidation, particularly of the fine material and rock surfaces.

5. Content of volatiles or volatile producing species (H_2O, SO_3, CO_2, NO_2).

6. Determination of the selected minor and trace element contents. (a) Primordial radionuclides: K with a sensitivity of at least 0.05 percent; U and Th with a sensitivity of at least 1 ppm. (b) Selected minor and trace elements (e.g., C, N, F, P, S, Cl, Ti, Ni, As, rare earth elements, Bi, Cu, Rb, Sr).

7. Measurement of physical properties (magnetic, and, in the case of fines, density and size distribution, and rheological properties). [pp. 44-45]

[1.5] The establishment of the nature and chronology of the major surface-forming processes should include determination of:

1. Cosmic-ray exposure ages of soil and rock materials for both long and short time scales.

2. Crystallization ages of igneous rocks, recrystallization ages of metamorphic rocks, and depositional ages of sedimentary rocks. [p. 44]

[1.6] The distribution and abundance of the volatile H_2O and CO_2 in the martian regolith should be determined to a depth of 2 m with an accuracy of 10 percent of the concentration and a sensitivity of detection of 0.1 percent. The surface temperature and temperature gradient should be measured. [p. 45]

[1.7] Evidence for the existence of life in the past or any information relative to the conditions under which it might evolve, are required to assess the biological potential of Mars. Among the measurements of the martian surface material that address this objective are the following:

a. A complete chemical analysis including all the principal chemical elements (those present in amounts greater than 0.5 percent by atom) as well as those of special biological significance (C, N, Na, P, S, Cl) with a sensitivity of at least 100 ppm;

b. A determination of the oxidation state of the sample and of the pH of water in equilibrium with it;

c. The quantitative determination of the function of depth;

d. The determination of the water-soluble constituents of the sample;

e. The determination of the major anions and cations present if the sample is exposed to water at various pH from 5 to 9;

f. The determination of the amounts of reduced carbon present with a sensitivity of 10 ppb;

g. The identification of the major mineral phases present;

h. The extensive search for possible fossil forms in martian soils and rocks. [p. 45]

[1.8] It is obvious that many of these measurements have pertinence to other than the biology oriented objectives of martian exploration.

Among the measurements that address the role of the environment of the martian samples and their ability to support life are

a. Establishment of the radiation environment at the surface of Mars, including electrons above 1 MeV and photons above 10 eV and

b. Determination of the amounts of minor constituents of the atmosphere (e.g., CH_4) that may reflect the existence of conditions someplace on Mars more favorable to the development of life than were found by Viking. [p. 46]

[1.9] Establishment of the interaction of the surface material with the atmosphere and its radiation environment should include the following investigations in addition to the specific analyses of surface material given above:

1. Reactivity of fine material with the constituents of the atmosphere (e.g., solubility in water, absorptive properties for CO_2, H_2O, CO, or O_2).

2. Noble-gas contents and isotopic composition of atmosphere and soil to a precision of better than O.5 percent for all major isotopes.

3. Determine the composition of the martian atmosphere at the surface over an annual cycle.

4. Precise determination of oxygen, nitrogen, carbon, and hydrogen isotope ratios in selected components of martian surface material and atmosphere. [p. 46]

[1.10] The circulation of the atmosphere of Mars provides the closest analogy to that of the Earth in the solar system, and it therefore serves as an ideal test site for dynamical and climatic theories developed for the Earth. Mechanical and thermal effects of topography on circulation, baroclinic instability, forcing and propagation of tides, generation of dust storms, and long-term climatic variations represent specific topics relevant to both the Earth and Mars. Neither the Viking landers nor Mariner orbiters have provided adequate information to define the global circulation pattern. Atmospheric temperature measurements with a resolution of roughly 5 degrees in latitude, 30 degrees in longitude, and 5 km in altitude between the surface and at least 30 km are needed. This goal could be achieved using a downward viewing infrared sounder in a low-altitude, circular, polar orbit. Much more detailed knowledge of atmospheric waves, including tides and Rossby waves, of the hydrological cycle, of regional meteorology, of the role of dust in the general circulation, and of winds above the boundary layer, is also needed. These problems could be addressed using about four ground-based stations with lifetimes exceeding one martian year and spaced between high latitudes and tropical regions. These stations should be sited to provide at least one triangular network with roughly 1000 km sides, and each station should measure pressure, temperature, relative humidity, atmospheric opacity, and wind velocity. The benefits to be derived from simultaneous measurements from the orbiter and the ground station network should be determined and assessed. [pp. 46–47]

[1.11] Determination of the internal structure of Mars, including the thickness of a crust and the existence and size of a core, and measurement of the location, size, and temporal dependence of martian seismic events is an objective of the highest importance. The level of martian seismicity, however, has not been established by the Viking seismology experiment. The possibility cannot be excluded on the basis of currently available data that the seismicity level may be substantially below the upper bound set by the Viking 2 seismometer results and/or that the absorption characteristics of the martian interior may be comparable with or enhanced over those of the earth's mantle. In such an eventuality, the number of seismic signals recordable on the martian surface from distances of greater than 1000 km may be very few.

In spite of this uncertainty, which has been recognized in assessing the relative priority of the determination of internal structure and dynamics as a major scientific objective for Mars exploration, we regard the likelihood of detectable natural seismic events as sufficiently high to recommend that a passive seismic network be established on the martian surface. Such a network should consist of at least three stations with broadband sensors, each with a sensitivity at least 100 times improved over the Viking seismometers, spaced approximately 1000 km apart and operating simultaneously for a period of at least one year. [p. 47]

[1.12] Accurate determination of the moment of inertia of Mars, a valuable constraint on internal structure, requires measurement of the martian precessional constant. This measurement can be made from the long-term tracking of one or more landed transmitters, an experiment that may also yield information on the existence of a martian Chandler wobble and on other polar motions. Combined mapping of gravity and topography will allow global extrapolation of locally derived seismic structure and will address the question of martian isostasy as a function of space and time. [p. 47]

[1.13] Determination of the character of the martian magnetic field and elucidation of the nature of the planet's interaction with the solar wind and the structure and dynamics of the upper atmosphere are essential objectives of continuing Mars study. . . .

Measurements sufficient to separate an internal, global-scale magnetic field from the solar-wind-induced field and to establish the presence of an internal field having a surface intensity approaching 10^{-5} G should be carried out. Confident separation of internal global and regional fields from the induced external components would be facilitated by simultaneous measurements of both the plasma and the magnetic field as well as measurements in the free-streaming solar wind. [p. 48]

[1.14] The interaction between the solar wind and Mars' upper atmosphere presents a host of problems that are fundamental to our understanding of both Mars and of planetary atmospheres in general. Among the major issues are the physical processes that produce mass exchange between the atmosphere and the solar-wind flow and the atmospheric mass-loss (or gain) rates that result; these escape processes are essential to our understanding of the evolution of Mars' atmosphere.

Characterization of the Mars solar-wind interaction will require establishing the distribution of neutral atmospheric constituents, as well as the ionized plasma and charged-particle distributions from both the solar wind and the atmosphere separately. These should be carried out both in the dayside interaction region, near an altitude of 300 km, and in the nightside, downstream magnetosphere, or wake region ranging to several Mars radii. In addition, the fluxes of energetic particles that may be accelerated by the Mars solar-wind interaction should be established. [p. 48]

[1.15] Potassium, thorium, and uranium should be determined to a sensitivity comparable with the levels in Apollo 11 basalts, and the following elements with an accuracy of 10 percent at the indicated concentrations: Fe, 1 percent; Ti, 0.5 percent; Si, 5.0 percent; O, 5.0 percent; Mg, 4.0 percent; H, 1.0 percent. The measurement of Al, Ca, Na, Mn, and Ni would be highly desirable. [p. 49]

[1.16] The diversity of the martian surface, as well as the wide range of environmental conditions and our ignorance of some of the key processes active on the martian surface, compel us to the view that the scientific objectives will best be met by exploring broad areas that exhibit the effects of distinctive processes that have influenced martian involution and by the intensive study of an intelligently selected suite of martian samples returned to Earth. The selection of returned materials should be based on our understanding of the global and local diversity of martian terrains and environments. [p. 49]

[1.17] To understand the current and past processes operating both at and near the surface of Mars, it is essential to explore the diversity of martian terrains that are apparent on both global and local scales. We therefore recommend that detailed exploration, on both global and local scales, of the diverse environments of Mars for purposes of understanding surface, near-surface, and atmospheric processes is a worthy goal in its own right and should be accomplished within the next decade. To this end, intensive local investigations in selected areas of 10 to 100 km in extent should be carried out, and, in addition, measurements at single points of extreme planetary environments should if possible be exploited. These local investigations should explore terrain and sample diversity with a wide range of chemical, mineralogical, and physical techniques. Both the analytical techniques and the manipulative skills of the experimental devices should be much advanced from those used on Viking, but without attempting to duplicate an Earth laboratory. Several science objectives requiring global-scale investigations can be accomplished with orbiters. Geochemical and geophysical mapping and atmospheric temperature soundings should if possible be carried out over the entire planet with spatial resolution compatible with science objectives. We emphasize that geochemical and geophysical mapping experiments must provide results that are clearly interpretable in terms of fundamental planetary characteristics and processes. [p. 49]

[1.18] Geochemical and geophysical mapping and atmospheric temperature soundings should be carried out over the entire planet with spatial resolution compatible with science objectives. We emphasize that geochemical and geophysical mapping experiments must provide results that are clearly interpretable in terms of fundamental planetary characteristics and processes. In addition, temperature sounding should cover the full diurnal and seasonal cycles. Investigation of Mars' magnetic field and atmospheric interaction with the solar wind requires both dayside and nightside measurements. [p. 50]

[1.19] The Space Science Board (see *Opportunities and Choices in Space Science*, 1974, National Academy of Sciences, Washington, D.C., 1975, p. 19) has previously recommended for Mars that the "long-term objectives of exobiology and surface chemistry investigation are best served by the return of an unsterilized surface sample to Earth" and further recommended that Mars sample return be adopted as a long-term goal. The Committee has thoroughly reconsidered this matter and concluded that understanding of the basic physical-chemical mechanisms that govern the surface of Mars can only be obtained by sophisticated and interactive analytical investigations. The return of martian surface and subsurface samples to Earth laboratories will allow the full range of the most sophisticated analytical techniques to be applied for the study of chronology, elemental and isotopic chemistry, mineralogy, and petrology and for the search for current and fossil life. In addition, such samples will be available to future scientists for study with improved techniques or with wholly new concepts compared with those available at the time the sample return mission was designed. We therefore reaffirm our view that the return of unsterilized surface and subsurface samples to Earth is a major technique for the exploration of Mars. Samples of distinctive materials, including rocks and fines, should be selected from an area of at least 2 m^2, based on visual inspection and major elemental analyses at the landing site. Materials should be selected that reflect the diversity of the local environment and the processes of broader planetary evolution. Samples should be returned to Earth in a manner that preserves their integrity and that is free from terrestrial contamination. [p. 50]

[1.20] With regard to the role of life-seeking experiments in the future exploration of Mars, COMPLEX is in accord with the general views expressed by CPBCE [Committee on Planetary Biology and Chemical Evolution, a former Committee of the Space Studies Board]. Based on the goal of understanding how the appearance of life in the solar system is related to the chemical history of the solar system, COMPLEX has formulated a strategy for future Mars exploration on the following premises:

1. Characterization of the chemical composition and physical state of materials on the surface and below the surface and the interaction of these materials with the atmosphere and sunlight are of basic importance to understanding the biologic potential of the planet.
2. The abundance and distribution of carbon compounds and water (including liquid water) in different materials is of significance.
3. The direct search for the study of chemical effects that relate to metabolic activity in martian materials and the intensive search for possible martian fossils should be carried out on unsterilized material returned to Earth without contaminating them with terrestrial materials.
4. Substantial attention and sensitivity toward the biologic potential of the martian environment should be associated with the in situ chemical and physical characterization of Mars without directing specific efforts towards active life-seeking experiments. [pp. 53–54]

[1.21] The CPBCE report distinguished between biologically relevant experiments that should be conducted remotely on the surface of Mars in an ensuing mission or missions and those that should be conducted on samples returned to Earth. For the former, it recommended analyses on samples of those characteristics that would constitute items of paramount importance to present or past biology and to organic chemical evolution, namely, the presence of reduced carbon, and the isotopic state of carbon, the amount and state of water, the presence of water-soluble electrolytes, and the existence of nonequilibrium gas compositions. It recommended that specific "life-seeking" metabolic-type experiments not be conducted remotely on the martian surface, but that they only be conducted on unsterilized samples returned to Earth. [p. 53]

2. 1990

The Search for Life's Origins: Progress and Future Directions in Planetary Biology and Chemical Evolution (NRC, CPBCE, 1990)

[The Committee on Planetary Biology and Chemical Evolution] recommends studies to:

[2.1] Conduct chemical, isotopic, mineralogical, sedimentological, and paleontological studies of martian surface materials at sites where there is evidence of hydrologic activity in any early clement epoch, through in situ determinations and through analysis of returned samples; of primary interest are sites in the channel networks and outflow plains; highest priority is assigned to sites where there is evidence suggestive of water-lain sediments on the floors of canyons as in the Valles Marineris system, particularly Hebes and Candor chasmata. [p. 124]

[2.2] Reconstruct the history of liquid water and its interactions with surface materials on Mars through photogeologic studies, space-based spectral reflectivity measurements, in situ measurements, and analysis of returned samples. [p. 124]

3. 1990

1990 Update to Strategy for Exploration of the Inner Planets (NRC, COMPLEX, 1990)

[3.1] The importance of the scientific objectives of study of the martian atmosphere, interior, magnetic field, and global properties should be given equal priority with the objective of intensive study of local areas. [p. 5]

[3.2] The geochemical, isotopic, and paleontological study of martian surface material for evidence of previous living material should be a prime objective of future in situ and sample return missions. [p. 5]

[3.3] Consistent with the SSB [Space Studies Board] report *The Search for Life's Origins: Progress and Future Directions in Planetary Biology and Chemical Evolution* (National Academy Press, Washington, D.C., 1990), the committee endorses the continued search for evidence of past life and biochemical evolution on Mars, as well as the continuing study of the history of water on Mars. [p. 21]

4. 1994

An Integrated Strategy for the Planetary Sciences: 1995-2010 (NRC, COMPLEX, 1994)

[4.1] Two kinds of precise positional measurements provide information on internal structure and dynamics. The first is a very accurate determination of the spin angular-momentum vector of a planet (both amplitude and direction) to monitor length-of-day changes, nutation, and precession. In some circumstances, such measurements can allow determination of the planet's first-order interior structure and whether the planet has a liquid core, as well as the nature of core-mantle coupling; this has been done for the Moon and could be done for Mars and Mercury.

The second type of measurement, which is regional and is similar to that made possible by the Global Positioning System on Earth, can lead to the detection of small relative crustal movements (of the order of 1 cm/yr or, possibly, 1 mm/yr in the future). Such measurements could provide interesting new information for a planet with suspected active tectonism, such as Venus and possibly Mars. [p. 88]

[4.2] Sample return may remain the only viable way of determining chronologies, but it should be emphasized that determination of even relatively imprecise ages can be very valuable in some cases. The flexibility, affordability, and feasibility of achieving many of these goals would be greatly enhanced by development of even crude dating techniques that could be placed aboard landed science packages. [pp. 98–99]

[4.3] A better understanding of the present climate of Mars inevitably depends also on understanding its present general circulation—the means by which heat, carbon dioxide, water vapor, and dust are transported. General circulation model simulations have shown that the dramatic martian seasonal surface-pressure variation, measured by the Viking landers, has two comparable components—one due to seasonal exchange with the polar caps and the other due to redistribution of atmospheric mass by the large-scale circulation. The modeling shows that a quantitative understanding of the seasonal CO_2 cycle and of the intimately linked cycles of dust and water requires knowledge of the large-scale seasonally varying pattern of atmospheric pressure and the closely related surface wind pattern responsible for raising and redistributing dust. Orbiters can determine the atmospheric temperature field and the dust and water loading but cannot measure the surface pressure with sufficient accuracy, and the pressure is a crucial dynamic boundary condition. Conversely, information on the surface pressure without data on the thermal field through the interior of the atmosphere is incomplete information. Ideally, the orbiter and lander measurements should be conducted simultaneously, because together they permit the construction of the full three-dimensional circulation. It has long been recognized that an orbiter, together with at least 15 or 20 surface stations, is required to achieve a good characterization of the system. [pp. 128–129]

[4.4] To resolve the issue of whether or not Mars had an early warm climate, the processes that created the observed channels and valley networks need to be elucidated, and the climatic implications of the processes need to be determined. More specific knowledge of the water budget in the crust of Mars and more accurate determinations of isotopic abundances, for example, the D/H abundance ratios, in the atmospheres of Venus and Mars will help to resolve this issue. [p. 131]

[4.5] The martian atmosphere is a high-priority region for study. It presents questions of climate variability, atmospheric origin, chemical stability, and atmospheric dynamics. Many of these questions are of particular interest among a broad community because Mars is similar enough to Earth to allow scientifically useful comparisons. Particular emphasis should be placed on long-term monitoring of dynamical behavior with good spatial resolution, such as can be performed by an orbiter. Surface meteorological stations, preferably accompanied by use of an orbiter, are the next step. Eventually, subsurface volatile reservoirs will need to be investigated to reach an understanding of atmospheric and climate history. [p. 135]

[4.6] At Mars, it is important to gain a first-order understanding of how the solar wind interacts with this planet and to begin the study of martian aeronomy. [p. 172]

[4.7] Mars is a marvelous place to study the processes that control atmospheric dynamics on terrestrial planets. Significant progress can be made through the deployment of a long-lived global network of surface meteorological stations. These outposts should provide essential data on the daily weather and, when combined with simultaneous sounding from orbit, will lead to much-improved general circulation models. These stations should also be used to determine the seasonal cycles of carbon dioxide, water, and dust and thereby learn something about how the layered martian polar sediments are deposited . . . .

One of the outstanding unknowns in geophysics concerns the internal structures of planets. This subject has profound ramifications for studies of origins and surface geology, since differentiation provides heat to mix the original materials and to shape later events. The easiest way to probe beneath Mars's surface is with a set of widely spaced seismometers that could be placed aboard the meteorological stations described above. These same stations should carry sophisticated geochemical laboratories to assay local materials. [p. 192]

5. 1995

An Exobiological Strategy for Mars Exploration (NASA, 1995)

[5.1] High-resolution imaging of selected sites by means of mid-IR mapping spectrometry is needed to identify surface expressions of aqueous mineralization, such as hydrothermal systems or spring deposits. [p. 54]

[5.2] We recommend establishment of a sequence of landed missions, beginning with development of a geochemically oriented payload capable of regional chemical and mineralogical analyses, oxidant identification, and volatile-element detection. This payload should be dispatched to a geologically diverse range of sites, which would be identified by means of high-resolution orbital data. This series of landed missions would lead in turn to identification of a limited number of sites of well-defined exobiological interest to which would then be dispatched a more exobiologically focused payload incorporating molecular and isotopic analysis of crustal volatiles. This phase of exploration would also involve high resolution local imagery aimed at assessing local lithologies for their potential to preserve fossils or to harbor extant life. [p. 54]

[5.3] Positive results for either preserved organic matter, potentially fossiliferous rocks, or habitats suggestive of extant life would then require deployment of highly focused experiments designed to test for modes of prebiotic chemistry, the presence of fossils, or evidence for metabolic activity, respectively. [p. 54]

[5.4] Landed missions should possess the mobility necessary to generate regional rather than purely local data. Such mobility will allow access to sites that may be virtually unreachable by fixed landers for landing safety considerations. [p. 54]

[5.5] Many of the techniques in geochemistry and paleontology that are used in exobiology-related studies on Earth do not lend themselves to field applications. Of particular relevance here is the difficulty that may be anticipated in conclusively identifying fossils of past life on Mars without returning a sample. Furthermore any positive signal from a robotic life-detection experiment would obviously demand confirmation in a terrestrial laboratory. For these reasons we recommend sample return as a key part of the long-range exobiology mission strategy. Clearly a sample return would follow after a series of surface lander and rover missions had analyzed samples from sites that had been identified as of particular interest. [p. 54]

6. 1996

"Scientific Assessment of NASA's Mars Sample-Return Mission Options" (NRC, COMPLEX, 1996)

[6.1] . . . [I]f the single objective of sample-return missions is to resolve the question of life on Mars, then highly successful missions could be characterized as failures if they do not return microfossils or living organisms. Therefore, justification of missions in terms of their bearing on the question of martian life alone would be a disservice to the scientific community and to the public, and would have a detrimental impact on the potential scientific results for exobiology and the other planetary science disciplines. Consequently, NASA should focus its Mars program, and sample-return missions in particular, on the more comprehensive goal of understanding Mars as a possible abode of life, a goal that is fully compatible with previous recommendations. [p. 2]

[6.2] [COMPLEX] is guardedly optimistic that NASA's current planning for Mars sample return missions will be consistent with the priorities outlined in past NRC [National Research Council] reports, provided that NASA takes into account the issues discussed above, as summarized here:

1. Formulate a program of Mars sample-return missions in the context of recent developments in the planetary, life, and astronomical sciences and directed toward the comprehensive goal of understanding Mars as a possible abode of life.
2. Incorporate previously developed strategies for determining "prebiotic" chemical evolution into the Mars sample-return program.
3. Maintain adaptability and flexibility in the Mars sample-return program to take into account possible new discoveries from ongoing Mars missions.
4. Ensure that the global reconnaissance of Mars is implemented as early as possible.
5. Ensure that sites and samples are selected that are consistent with established strategies for exobiology and martian exploration.
6. To understand the results from each mission and to provide input for the planning of ongoing missions during the entire Mars exploration program, there must be an adequate, ongoing data-analysis program.

7. Ensure that sample handling strategies, including planetary protection issues, are judiciously formulated and implemented.

8. Develop the capability for achieving long-range (tens of kilometers) mobility and high-precision landing.

9. Develop a broad suite of capable, miniature instruments for in situ measurements of surface properties relevant to exobiology and general martian exploration.

10. Develop the criteria to enable the unambiguous identification of biotic signatures.

11. Increase the rate of collection of antarctic meteorites relevant to Mars by, for example, increasing the efficiency of field collection procedures. [p. 5]

7. 1996

"The Search for Evidence of Life on Mars" (McCleese Report) (NASA, Mars Expeditions Strategy Group, 1996)

[7.1] The members of the Mars strategy group recommend that the search for life on Mars should be directed at locating and investigating, in detail, those environments on the planet which were potentially most favorable to the emergence (and persistence) of life: ancient ground water environments; ancient surface water environments; and modern ground water environments.

[7.2] We urge strongly that the investigation strategy emphasize sampling at diverse sites. It is specifically recommended that the implementation of the program of exploration of Mars be aimed at the study of a range of ancient and modern aqueous environments. These environments may be accessed by exploring the ejecta of young impact craters, by investigating material accumulated in outflow channels, and by coring.

[7.3] In-situ studies conducted on the surface of Mars are essential to our learning more about Martian environments and for selecting the best samples for collection. However, for the next 10 years or more, the essential analyses of selected samples must be done in laboratories on Earth . . . "high precision" (i.e., sophisticated, state-of-the-art) analytical techniques must be used, such as those found in only the most advanced laboratories here on Earth.

[7.4] We also believe that to achieve widely accepted confirmation of Martian life, all three of the following must be clearly identified and shown to be spatially and temporally correlated within rock samples: 1) organic chemical signatures that are indicative of life, 2) morphological fossils (or living organisms), 3) supporting geochemical and/or mineralogical evidence (e.g., clearly biogenic isotopic fractionation patterns, or the presence of unequivocal biominerals). These characteristics can not be properly evaluated without the return of a variety of Martian samples to Earth for interdisciplinary study in appropriate laboratories.

[7.5] Precursor orbital information must be obtained, as well, to select the best sites for surface studies. We can already say with reasonable certainty, however, that the ancient highlands represent a region of great potential, and that at least the initial focused studies should be performed there. Maps of surface mineralogy will be needed to enhance investigations within the highlands and enable searches elsewhere. This work begins with the launch of the Mars Global Surveyor (MGS) later this year. Additional measurements from orbit at higher spatial resolution are essential to identify productive sites (e.g., regions containing carbonates) at scales accessible by surface rovers. In addition, instruments capable of identifying near-surface water, water bound in rocks, and subsurface ice, would greatly accelerate and make more efficient our search for environments suitable for life.

[7.6] For ancient ground water environments, a sample return mission can occur relatively soon, since the necessary precursor information for site selection is already available from existing orbital photogeologic data, including Mariner 9 and Viking imagery, or will be provided by Mars Surveyor orbiters in '96, '98 and '01.

[7.7] For ancient surface water environments, orbital and surface exploration/characterization should precede sample return because identification of extensive areas of carbonates and evaporites is highly desirable. This implies the use of advanced orbital and in-situ instruments for mineral characterization.

[7.8] Sample return missions will retrieve the most productive samples if they are supported by extensive searches, analyses and collections performed by sophisticated rovers. These should be capable of ranges of 10's of kilometers in order to explore geologically diverse sites. The specific samples to be returned to Earth would be selected using criteria that increase the probability of finding direct evidence of life as well as the geological context, age and climatic environment in which the materials were formed.

8. 1996

"Scientific Assessment of NASA's Solar System Exporation Roadmap" (NRC, COMPLEX, 1996)

[8.1] COMPLEX has attached very high priority to a better understanding of martian atmospheric circulation as the key component of the climate system and for comparative studies of atmospheric dynamics. Yet, this Roadmap campaign does not effectively address this key objective for Mars. [p. 4]

9. 1998

"Assessment of NASA's Mars Exploration Architecture" (NRC, COMPLEX, 1998)

[9.1] . . . [A]n appropriate focus for NASA's Mars program is the comprehensive goal of understanding Mars as a possible abode of past or present life. [p. 10]

[9.2] To the extent possible, information must be obtained on the global martian environment in order to understand the events in the history of the martian samples and of the planet in general. [p. 2]

10. 2000

NASA Strategic Plan 2000 (NASA, 2000)

Only items relevant to Mars exploration are listed.

[10.1] Objectives

- Learn how galaxies, stars, and planets form, interact, and evolve
- Look for signs of life in other planetary systems
- Understand the formation and evolution of the solar system and the Earth within it
- Probe the evolution of life on Earth, and determine if life exists elsewhere in the solar system
- Understand our changing Sun and its effects throughout the solar system
- Investigate the composition, evolution, and resources of Mars, the Moon, and small bodies [p. 18]

[10.2] Near-term Plans (2000–2005)

- Investigate Saturn, its rings, and moon Titan. Analyze the structure and composition of comets, understand the history of Mars, and return dust and solar wind samples
- Conduct laboratory and field research on the origin of life on Earth (Astrobiology Initiative), and search for water on Mars
- Obtain images of the Earth's magnetosphere during geomagnetic storms, search for evidence of water on Mars, and characterize the number and orbits of Near Earth Objects
- Explore the surface and atmosphere of Mars, survey the structure and composition of asteroids, and investigate the composition and structure of comets [p. 18]

[10.3] Mid-term Plans (2006–2011)

- Learn about formation of the rocky planets, investigate the nature of the early solar system by returning a sample from a comet, and continue exploration of Mars
- Continue research on life on Earth and potential biological history of Mars, and search for liquid water ocean on Jupiter's moon, Europa
- Investigate selected sites on Mars in detail
- Continue exploration of Mars, ascertain the presence of a liquid water ocean on Europa, and return a sample from a comet nucleus [p. 19]

[10.4] Long-term Plans (2012–2025)

- Complete reconnaissance of the Solar System by flying by Pluto, studying Neptune and its satellite Triton, and conducting advanced studies of Mars
- Search for evidence of biological activity on Europa, study the prebiotic chemistry of Saturn's moon Titan, and explore promising solar system targets to search for evidence of past or present life. Integrate solar system findings with the search for life in other planetary systems
- Continue exploration of the inner solar system in support of possible human exploration [p. 19]

11. 2000

"Mars Exploration Program: Scientific Goals, Objectives, Investigations, and Priorities" (NASA, MEPAG, 2001)

Extracted from the December 2000 edition of the MEPAG report, "Mars Exploration Program: Scientific Goals, Objectives, Investigations, and Priorities," edited by R. Greeley. An exhaustive discussion of Mars science priorities, the MEPAG report is at once valuable and frustrating. It is valuable because it is one of the few committee studies that makes any attempt to prioritize science objectives. It is frustrating because it is so ambitious and inclusive as to be unrealistic. Detailed discussions have been removed in the extracted material that follows to make the list short enough to be included in this appendix. The report is organized into broad Goals, and each goal into narrower Objectives. Quoting from the report:

> Within each objective, the investigations are listed [in priority order] as determined within each discipline. There was no attempt to synthesize the overall set of investigations, but it was recognized that synergy among the various goals and objectives could alter the priorities in an overall strategy. Completion of all the investigations will require decades of effort. It is recognized that many investigations will never be truly complete (even if they have a high priority) and that evaluations of missions should be based on how well the investigations are addressed. While priorities should influence the sequence in which the investigations are conducted, it is not intended that they be done serially, as many other factors come into play in the overall Mars Program . . . . [pp. 1–2]

This section groups all the top-priority (No. 1) investigations listed, across objectives, under a Category 1. The same is done with Categories 2, 3, . . . 10. MEPAG would protest, with some justice, that this is a pointless and misleading exercise. Constructing a top-priority category with contributions from all the objectives makes the assumption that all "objectives" are equally important. This, of course, is not the case, but MEPAG, like every other study group, refused to consider the relative importance of the objectives and the disciplines they reside in. This method of grouping also penalizes those objectives whose representatives recognized many needed investigations, which relegates most of the investigations to high-numbered (implied low-priority) categories.

Nonetheless, the categories, at least the low-numbered ones, do crudely express MEPAG's priorities, which gives the categories some value. For more detail and fairness, the reader is directed to the original MEPAG report.

11.1, Category 1

[11.1.1] Map the 3-dimensional distribution of water in all its forms. . . . Requires global mapping by remote sensing and, if possible, seasonal changes in near-surface water budgets. [p. 2]

[11.1.2] Determine the locations of sedimentary deposits formed by ancient and recent surface and subsurface hydrological processes. . . . Requires global geomorphic and mineral mapping, followed by the in situ "ground truth" for orbital data and to identify sites for sample return. [p. 4]

[11.1.3] Search for complex organic molecules in rocks and soils. . . . Requires studies of modern aqueous environments and aqueous paleoenvironments preserved in ancient sedimentary rocks. Targets for in situ studies must be first identified from orbit, then mobile platforms (rovers), and returned samples. [p. 6]

[11.1.4] Determine the processes controlling the present distributions of water, carbon dioxide and dust. . . . Requires global mapping and then landed observations on daily and seasonal time scales. [p. 7]

[11.1.5] Find physical and chemical records of past climates. . . . Requires remote sensing of stratigraphy and aqueous weathering products, landed exploration, and returned samples. [p. 10]

[11.1.6] Determine the present state, distribution and cycling of water on Mars. . . . Requires global observations using geophysical sounding and neutron spectroscopy, coupled with measurements from landers, rovers, and the subsurface. [p. 12]

[11.1.7] Characterize the configuration of Mars' interior. . . . Requires orbital and lander data. [p. 17]

[11.1.8] Determine the radiation environment at the Martian surface and the shielding properties of the Martian atmosphere [HEDS]. . . . Requires simultaneous monitoring of the radiation in Mars' orbit and at the surface, including the ability to determine the directionality of the neutrons at the surface. [p. 18]

11.2, Category 2

[11.2.1] Carry out in situ exploration of possible liquid water in the subsurface. . . . Requires drilling to km depths and instruments to detect water in all forms, CO_2 clathrate, and to analyze rocks, soils and ices for organic compounds or to detect life. [p. 2]

[11.2.2] Search for Martian fossils. . . . Requires orbital mapping, in situ analysis, and sample returns. [p. 5]

[11.2.3] Determine the changes in crustal and atmospheric inventories of organic carbon through time. . . . [T]his objective is posed in a historical way that requires a stratigraphic framework (established through geological mapping) and returned samples. [p. 7]

[11.2.4] Determine the present-day stable isotopic and noble gas composition of the present-day bulk atmosphere. [p. 9]

[11.2.5] Characterize history of stratigraphic records of climate change at the polar layered deposits, the residual ice caps. . . . Requires orbital, in situ observations and returned samples. [p. 11]

[11.2.6] Evaluate sedimentary processes and their evolution through time, up to and including the present. . . . Requires knowledge of the age, sequence, lithology and composition of sedimentary rocks (including chemical deposits), as well as the rates, durations, environmental conditions, and mechanics of weathering, cementation, and transport processes. [p. 12]

[11.2.7] Determine the history of the magnetic field. . . . Requires orbiter in eccentric orbit or low-altitude platform. [p. 17]

[11.2.7] Characterize the chemical and biological properties of the soil and dust [HEDS]. . . . The requirements can and may have to be met through sample studies on Earth. Earth sample return provides significant benefits to HEDS technology development programs. [p. 19]

11.3, Category 3

[11.3.1] Explore high priority candidate sites (i.e., those that provide access to near-surface liquid water) for evidence of extant (active or dormant) life. . . . Requires in situ life experiments on subsurface materials and laboratory analysis of returned samples. [p. 3]

[11.3.2] Determine the timing and duration of hydrologic activity. . . . Requires the development of stratigraphic (age) framework, in situ measurements, and sample returns from key sites for radiometric dating. [p. 6]

[11.3.3] Determine long-term trends in the present climate. [p. 9]

[11.3.4] Calibrate the cratering record and absolute ages for Mars. . . . Requires absolute ages on returned rock (not soil) samples of known crater ages. [p. 13]

[11.3.5] Determine the chemical and thermal evolution of the planet. . . . Requires measurements from orbiter and lander. [p. 18]

[11.3.6] Understand the distribution of accessible water in soils, regolith, and Martian groundwater systems [HEDS]. . . . Requires geophysical investigations and subsurface drilling and in situ sample analysis. [p. 20]

11.4, Category 4

[11.4.1] Determine the array of potential energy sources to sustain biological processes. . . . Requires orbital mapping and in situ investigations. [p. 3]

[11.4.2] Determine the rates of escape of key species from the Martian atmosphere, and their correlation with solar variability and lower atmosphere phenomenon (e.g. dust storms). . . . Requires: Global orbiter observations of species (particularly H, O, CO, CO_2 and key isotopes) in the upper atmosphere, and monitoring their variability over multiple Martian years. [p. 9]

[11.4.3] Evaluate igneous processes and their evolution through time, including the present. . . . Requires global imaging, geologic mapping, techniques for distinguishing igneous and sedimentary rocks, evaluation of current activity from seismic monitoring, and returned samples. [p. 13]

[11.4.4] Measure atmospheric parameters and variations that affect atmospheric flight [HEDS]. . . . Requires instrumented aeroentry shells or aerostats. [p. 20]

11.5, Category 5

[11.5.1] Determine the nature and inventory of organic carbon in representative soils and ices of the Martian crust. . . . Requires in situ exploration and sample return. [p. 4]

[11.5.2] Search for micro-climates. . . . Requires global search for sites based on topography or changes in volatile distributions and surface properties (e.g., temperature or albedo). [p. 10]

[11.5.3] Characterize surface-atmosphere interactions on Mars, including polar, eolian, chemical, weathering, and mass-wasting processes. Interest here is in processes that have operated for the last million years as recorded in the upper 1 m to 1 km of geological materials. . . . Requires orbital remote sensing of surface and subsurface, and in situ measurements of sediments and atmospheric boundary layer processes. [p. 14]

[11.5.4] Determine electrical effects in the atmosphere [HEDS]. . . . Requires experiments on a lander. [p. 21]

11.6, Category 6

[11.6.1] Determine the distribution of oxidants and their correlation with organics. . . . Requires instruments to determine elemental chemistry and mineralogy. [p. 4]

[11.6.2] Determine the production and reaction rates of key photochemical species (O_3, H_2O_2, CO, OH, etc.) and their interaction with surface materials. [p. 10]

[11.6.3] Determine the large-scale vertical structure and chemical and mineralogical composition of the crust and its regional variations. This includes, for example, the structure and origin of hemispheric dichotomy. . . . Requires remote sensing and geophysical sounding from orbiters and surface systems, geologic mapping, in-situ analysis of mineralogy and composition of surface material, returned samples, and seismic monitoring. [p. 15]

[11.6.4] Measure the engineering properties of the Martian surface [HEDS]. . . . Requires in-situ measurements at selected sites. [p. 21]

11.7, Category 7

[11.7.1] Document the tectonic history of the Martian crust, including present activity. . . . Requires geologic mapping using global topographic data combined with high-resolution images, magnetic and gravity data, and seismic monitoring. [pp. 15–16]

[11.7.2] Determine the radiation shielding properties of Martian regolith [HEDS]. . . . Requires an understanding of the regolith composition, a lander with the ability to bury sensors at various depths up to a few meters. Some of the in situ measured properties may be verified with a returned sample. [p. 22]

11.8, Category 8

[11.8.1] Evaluate the distribution and intensity of impact and volcanic hydrothermal processes through time, up to and including the present. . . . Requires knowledge of the age and duration of the hydrothermal system, the heat source, and the isotopic and trace element chemistry and mineralogy of the materials deposited. [p. 16]

[11.8.2] Measure the ability of Martian soil to support plant life [HEDS]. . . . Requires in-situ measurements and process verification. [p. 22]

11.9, Category 9

[11.9.1] Characterize the topography, engineering properties, and other environmental characteristics of candidate outpost sites. Site certification for human outposts requires a set of data about the specific site that can best be performed by surface investigations [HEDS]. [p. 22]

11.10, Category 10

[11.10.1] Determine the fate of typical effluents from human activities (gases, biological materials) in the Martian surface environment [HEDS]. [p. 22]

REFERENCES

Mars Expeditions Strategy Group, National Aeronautics and Space Administration (NASA), "The Search for Evidence of Life on Mars," 1996, available online at <http://geology.asu.edu/~jfarmer/mccleese.htm>. Also available in National Aeronautics and Space Administration, *Science Planning for Exploring Mars*, JPL Publication 01-7, Jet Propulsion Laboratory, Pasadena, Calif., 2001.

NASA, *An Exobiological Strategy for Mars Exploration*, NASA, Washington, D.C., 1995.

NASA, *Strategic Plan 2000*, NASA, Washington, D.C., 2000.

NASA, Mars Exploration Payload Assessment Group (MEPAG), "Mars Exploration Program: Scientific Goals, Objectives, Investigations, and Priorities," December 2000, in *Science Planning for Exploring Mars*, JPL Publication 01-7, Jet Propulsion Laboratory, Pasadena, Calif., 2001.

NRC (National Research Council), COMPLEX (Committee on Planetary and Lunar Exploration), *Strategy for Exploration of the Inner Planets: 1977–1987*, National Academy Press, Washington, D.C., 1978.

NRC, Committee on Planetary Biology and Chemical Evolution (CPBCE), *The Search for Life's Origins: Progress and Future Directions in Planetary Biology and Chemical Evolution*, National Academy Press, Washington, D.C., 1990.

NRC, COMPLEX, *1990 Update to Strategy for Exploration of the Inner Planets*, National Academy Press, Washington, D.C., 1990.

NRC, COMPLEX, *An Integrated Strategy for the Planetary Sciences: 1995-2010*, National Academy Press, Washington, D.C., 1994.

NRC, COMPLEX, "Scientific Assessment of NASA's Mars Sample-Return Mission Options," letter report, Space Studies Board, NRC, Washington, D.C., 1996.

NRC, COMPLEX, "Scientific Assessment of NASA's Solar System Exploration Roadmap," letter report, Space Studies Board, NRC, Washington, D.C., 1996.

NRC, COMPLEX, "Assessment of NASA's Mars Exploration Architecture," letter report, Washington, D.C., 1998.

Appendix C

Acronyms

ALH	designation for meteorites collected in Allan Hills, Antarctica
APXS	Alpha Proton X-ray Spectrometer (on NASA Sojourner rover); Alpha Particle X-ray Spectrometer (on NASA Mars Exploration Rover)
ASI	Agenzia Spaziale Italiana (Italian space agency)
ASPERA	Energetic Neutral Atoms Analyzer (on ESA Mars Express spacecraft)
ATHENA	integrated science package carried by NASA Mars Exploration Rover mission
ATMIS	surface meteorology package (on European NetLander spacecraft)
CNES	Centre National d'Etudes Spatiales (French space agency)
COMPLEX	Committee on Planetary and Lunar Exploration (of the Space Studies Board)
CPBCE	Committee on Planetary Biology and Chemical Evolution (of the Space Science Board)
DLR	Deutsches Zentrum für Luft- und Raumfahrt (German space agency)
FMI	Finnish Meteorological Institute
GAP	gas-analysis package—Gas Chromatograph-Mass Spectrometer (on U.K. Beagle 2 spacecraft)
GRS	Gamma-Ray Spectrometer (on NASA Mars Odyssey spacecraft)
HEDS	Human Exploration and Development of Space
HRSC	High Resolution Stereo Camera (on ESA Mars Express spacecraft)
IMP	Imager for Mars Pathfinder (NASA)
IR	infrared
JPL	Jet Propulsion Laboratory
LREE	light rare-earth element
LST	local solar time

MARCI	Mars Color Imager (on NASA Mars Reconnaissance Orbiter mission)
MARIE	Mars Radiation Environment Experiment (on NASA Mars Odyssey orbiter spacecraft)
MaRS	Radio Science Experiments (on ESA Mars Express spacecraft)
MARSIS	Subsurface Sounding Radar/Altimeter (on ESA Mars Express spacecraft)
MCO	Mars Climate Orbiter (NASA mission, lost)
ME	ESA Mars Express mission
MEP	Mars Exploration Program (NASA)
MEPAG	Mars Exploration Payload Assessment Group (NASA)
MER	Mars Exploration Rover (NASA mission)
MGS	Mars Global Surveyor (NASA mission)
Mini-TES	Miniature Thermal Emission Spectrometer (on NASA Mars Exploration Rover)
MO	Mars Observer (NASA mission, lost)
MOC	Mars Orbiter Camera (on NASA Mars Global Surveyor spacecraft)
MOLA	Mars Orbiter Laser Altimeter (on NASA Mars Global Surveyor spacecraft)
MPL	Mars Polar Lander (NASA mission, lost)
MRO	Mars Reconnaissance Orbiter (NASA mission)
MSL	Mars Science Laboratory (NASA mission)
MSR	Mars sample return
MVACS	Mars Volatiles and Climate Surveyor (integrated payload on lost Mars Polar Lander)
NASA	National Aeronautics and Space Administration
NASDA	National Space Development Agency (Japan)
NRC	National Research Council
OMEGA	Infrared Mineralogical Mapping Spectrometer (on ESA Mars Express spacecraft)
PFS	Planetary Fourier Spectrometer (on ESA Mars Express spacecraft)
PI	principal investigator
PMIRR	Pressure-Modulator Infrared Radiometer (on lost NASA Mars Observer and Mars Climate Orbiter spacecraft)
PMIRR-MkII	Pressure-Modulator Infrared Radiometer (on NASA Mars Reconnaissance Orbiter mission)
SNC	subset of meteorites that derive from Mars (shergottites, nakhlites, chassignites)
SOFIA	Stratospheric Observatory for Infrared Astronomy (NASA/DLR airborne observatory)
SPICAM	Ultraviolet and Infrared Atmospheric Spectrometer (on ESA Mars Express spacecraft)
SSB	Space Studies Board (after 1988), Space Science Board (before 1988)
TES	Thermal Emission Spectrometer (on NASA Mars Global Surveyor spacecraft)
THEMIS	Thermal Emission Imaging System (on NASA Mars Odyssey Orbiter spacecraft)